Andrew Theocritus

Theocritus, Bion and Moschus

Andrew Theocritus

Theocritus, Bion and Moschus

ISBN/EAN: 9783743333543

Manufactured in Europe, USA, Canada, Australia, Japa

Cover: Foto ©berggeist007 / pixelio.de

Manufactured and distributed by brebook publishing software
(www.brebook.com)

Andrew Theocritus

Theocritus, Bion and Moschus

CONTENTS

LIFE OF THEOCRITUS
(*From Suidas*)

THEOCRITUS, the Chian. But there is another
Theocritus, the son of Praxagoras and Philinna (see
Epigram XXIII), or as some say of Simichus. (This
is plainly derived from the assumed name Simichidas
in Idyl VII.) He was a Syràcusan, or, as others say,
a Coan settled in Syracuse. He wrote the so-called
Bucolics in the Dorian dialect. Some attribute to
him the following works :—*The Proetidae, The Plea-
sures of Hope* ('Eλπίδες), *Hymns, The Heroines, Dirges,
Ditties, Elegies, Iambics, Epigrams.* Be it known
that there are three Bucolic poets : this Theocritus,
Moschus of Sicily, and Bion of Smyrna, from a village
called Phlossa.

LIFE OF THEOCRITUS
ΘΕΟΚΡΙΤΟΤ ΓΕΝΟΣ
Usually prefixed to the Idyls)

THEOCRITUS the Bucolic poet was a Syracusan by
extraction, and the son of Simichidas, as he says
himself, *Simichidas, pray whither through the noon
dost thou drag thy feet?* (Idyl VII). Some say that
this was an assumed name, for he seems to have been
snub-nosed (σιμός), and that his father was Praxagoras,
and his mother Philinna. He became the pupil of
Philetas and Asclepiades, of whom he speaks (Idyl
VII), and flourished about the time of Ptolemy Lagus.
He gained much fame for his skill in bucolic poetry.
According to some his original name was Moschus,
and Theocritus was a name later assumed.

AT the beginning of the third century before Christ, in the years just preceding those in which Theocritus wrote, the genius of Greece seemed to have lost her productive force. Nor would it have been strange if that force had really been exhausted. Greek poetry had hitherto enjoyed a peculiarly free development, each form of art succeeding each without break or pause, because each—epic, lyric, dithyramb, the drama—had responded to some new need of the state and of religion. Now in the years that followed the fall of Athens and the conquests of Macedonia, Greek religion and the Greek state had ceased to be themselves. Religion and the state had been the patrons of poetry; on their decline poetry seemed dead. There were no heroic kings, like those for whom epic minstrels had chanted. The cities could no longer welcome an Olympian winner with Pindaric hymns. There was no imperial Athens to fill the theatres with a crowd of citizens and strangers eager to listen to new

tragic masterpieces. There was no humorous
democracy to laugh at all the world, and at
itself, with Aristophanes. The very religion of
Sophocles and Aeschylus was debased. A
vulgar usurper had stripped the golden orna-
ments from Athene of the Parthenon. The
ancient faith in the protecting gods of Athens,
of Sparta, and of Thebes, had become a lax
readiness to bow down in the temple of any
Oriental Rimmon, of Serapis or Adonis. Greece
had turned her face, with Alexander of Macedon,
to the East ; Alexander had fallen, and Greece
had become little better than the western portion of a divided Oriental empire. The centre
of intellectual life had been removed from
Athens to Alexandria (*founded* 332 B.C.) The
new Greek cities of Egypt and Asia, and above
all Alexandria, seemed no cities at all to Greeks
who retained the pure Hellenic traditions.
Alexandria was thirty times larger than the
size assigned by Aristotle to a well-balanced
state. Austere spectators saw in Alexandria
an Eastern capital and mart, a place of harems
and bazaars, a home of tyrants, slaves, dreamers,
and pleasure-seekers. Thus a Greek of the
old school must have despaired of Greek poetry.
There was nothing (he would have said) to
evoke it ; no dawn of liberty could flush this
silent Memnon into song. The collectors,
critics, librarians of Alexandria could only pro-
duce literary imitations of the epic and the
hymn, or could at best write epigrams or
inscriptions for the statue of some alien and

luxurious god. Their critical activity in every field of literature was immense, their original genius sterile. In them the intellect of the Hellenes still faintly glowed, like embers on an altar that shed no light on the way. Yet over these embers the god poured once again the sacred oil, and from the dull mass leaped, like a many-coloured flame, the genius of THEOCRITUS.

To take delight in that genius, so human, so kindly, so musical in expression, requires, it may be said, no long preparation. The art of Theocritus scarcely needs to be illustrated by any description of the conditions among which it came to perfection. It is always impossible to analyse into its component parts the genius of a poet. But it is not impossible to detect some of the influences that worked on Theocritus. We can study his early 'environment'; the country scenes he knew, and the songs of the neatherds which he elevated into art. We can ascertain the nature of the demand for poetry in the chief cities and in the literary society of the time. As a result, we can understand the broad twofold division of the poems of Theocritus into rural and epic idyls, and with this we must rest contented.

It is useless to attempt a regular biography of Theocritus. Facts and dates are alike wanting, the ancient accounts (p. ix) are clearly based on his works, but it is by no means impossible to construct a 'legend' or romance of his life, by aid of his own verses, and of hints and

fragments which reach us from the past and
the present. The genius of Theocritus was so
steeped in the colours of human life, **he bore**
such true and full witness as to the **scenes and
men he knew, that life (always** essentially the
same) becomes in turn a witness to his veracity.
He was born in the midst **of** nature that,
through all the changes of things, has never
lost its sunny charm. The existence he loved
best to contemplate, that of southern shepherds, fishermen, rural people, remains what
it always has been in Sicily and in the isles
of Greece. The habits and the passions of his
countryfolk have not **altered, the** echoes of
their old love-songs still sound **among** the
pines, or by the sea-banks, where Theocritus
'watched the visionary flocks.'

Theocritus was probably born in an early
decade of the third century, or, according to
Couat, about 315 B.C., and was a native of Syra-
cuse, 'the greatest of Greek cities, the fairest
of all cities.' So Cicero calls it, describing the
four quarters that were encircled by its walls,
—each quarter as large as a town,—the
fountain Arethusa, the stately temples with
their doors of ivory and gold. On the fortunate dwellers in Syracuse, Cicero says, the
sun shone every day, and there was never a
morning so tempestuous but the sunlight conquered at last, and broke through the clouds.
That perennial sunlight still floods the poems
of Theocritus with its joyous glow. His birthplace was the proper home of an idyllic poet,

of one who, with all his enjoyment of the city
life of Greece, had yet been 'breathed on by
the rural Pan,' and best loved the sights and
sounds and fragrant air of the forests and the
coast. Thanks to the mountainous regions of
Sicily, to **Etna,** with her volcanic cliffs and
snow-fed streams, thanks also to the hills **of**
the interior, the populous island never lost the
charm of nature. Sicily **was not like the over-
crowded and** over-cultivated **Attica; among**
the Sicilian heights and by the coast were few
enclosed estates and narrow farms. The
character of the people, too, was attuned to
poetry. The Dorian settlers had kept alive the
magic of rivers, of pools where the Nereids
dance, and uplands haunted by Pan. This
popular poetry influenced the literary verse **of**
Sicily. The songs of Stesichorus, a minstrel
of the early period, and the little rural 'mimes'
or interludes of Sophron are lost, and we have
only fragments of Epicharmus. But it seems
certain that these poets, predecessors of Theocritus, liked to mingle with their own compo-
sition strains of rustic melody, *volks-lieder*,
ballads, love-songs, ditties, and dirges, such as
are still chanted by the peasants of Greece and
Italy. Thus in Syracuse and the other towns
of the coast, Theocritus would have always
before his eyes the spectacle of refined and
luxurious manners, and always in his ears
the babble of the Dorian women, while he **had** .
only to pass the gates, and wander through the
fens of Lysimeleia, by the brackish mere, or

ride into the hills, to find himself in the golden
world of pastoral. Thinking of his early years,
and of the education that nature gives the
poet, we can imagine him, like Callicles in Mr.
Arnold's poem, singing at the banquet of a
merchant or a general—

'With his head full of wine, and his hair crown'd,
  Touching his harp as the whim came on him,
  And praised and spoil'd by master and by guests,
  Almost as much as the new dancing girl.'

We can recover the world that met his eyes
and inspired his poems, though the dates of the
composition of these poems are unknown. **We**
can follow him, in fancy, as he breaks from
the revellers and wanders out into the night.
Wherever he turned his feet, he could find
such scenes as he has painted in the idyls. If
the moon rode high in heaven, as he passed
through the outlying gardens he might catch a
glimpse of some deserted girl shredding the
magical herbs into the burning brazier, and
sending upward to the 'lady Selene' the song
which was to charm her lover home. The
magical image melted in the burning, the herbs
smouldered, the tale of love was told, and
slowly the singer 'drew the quiet night into her
blood.' Her lay ended with a passage of
softened melancholy—

  'Do thou farewell, and turn thy steeds to
Ocean, lady, and my pain I will endure, even
as I have declared. Farewell, Selene beautiful;

farewell, ye other stars that follow the wheels of Night.'

A grammarian says that Theocritus borrowed this second idyl, the story of Simaetha, from a piece by Sophron. But he had no need to borrow from anything but the nature before his eyes. Ideas change so little among the Greek country people, and the hold of superstition is so strong, that betrayed girls even now sing to the Moon their prayer for pity and help. Theocritus himself could have added little passion to this incantation, still chanted in the moonlit nights of Greece :[1]

'Bright golden Moon, that now art near to thy setting, go thou and salute my lover, he that stole my love, and that kissed me, and said, " Never will I leave thee." And, lo, he has left me, like a field reaped and gleaned, like a church where no man comes to pray, like a city desolate. Therefore I would curse him, and yet again my heart fails **me for** tenderness, my heart is vexed within me, my spirit is moved with anguish. Nay, even so I will lay **my** curse **on** him, and let God do even as He will, with my pain and with my crying, with my flame, and mine imprecations.'

It is thus that the women of the islands, like the girl of Syracuse two thousand years ago, hope to lure back love or avenged love betrayed, and thus they 'win more ease from song than could be bought with gold.'

[1] This fragment is from the collection of M. Fauriel ; *Chants Populaires de la Grèce.*

In whatever direction the path of the Syracusan wanderer lay, he would find then, as he would find now in Sicily, some scene of the idyllic life, framed between the distant Etna and the sea. If he strayed in the faint blue of the summer dawn, through the fens to the shore, he might reach the wattled **cabin of** the two old **fisher-**men in the twenty-first idyl. There is nothing in Wordsworth more real, more full of the incommunicable sense of nature, rounding and softening the toilsome days of the aged and the poor, than the Theocritean poem of the Fisher-**man's Dream.** It is **as true** to nature as the statue of the naked **fisherman** in the Vatican. One cannot read these verses but the vision returns to one, of sandhills by the sea, **of a** low cabin roofed with grass, where fishing-rods of reed are leaning against the door, while **the** Mediterranean floats up her waves that fill the waste with sound. This nature, grey and still, seems in harmony with the wise content of old men whose days are waning on the limit of life, as they have all been spent by the desolate margin of the sea.

The twenty-first idyl is one of the rare poems of Theocritus that are not filled with the sunlight of Sicily, or of Egypt. The landscapes **he** prefers are often seen under the noonday **heat,** when shade is most pleasant to men. His shepherds invite each other to the shelter of oak-trees **or of** pines, where the dry fir-needles are strown, or where the feathered ferns make a luxurious ' couch more soft than sleep,'

or where the flowers bloom whose musical
names sing in the idyls.   Again, Theocritus
will sketch the bare beginnings of the hillside,
as in the third idyl, just where the olive-gar-
dens cease, and where the short grass of the
heights alternates with rocks, and thorns, and
aromatic **plants**.   None **of his** pictures seem
complete without the **presence of water**.   It
may be **but the** wells **that the maidenhair**
fringes, **or the babbling runnel of the fountain
of** the **Nereids**.   The shepherds **may** sing of
Crathon, or Sybaris, or Himeras, waters so
sweet that they seem to flow with milk and honey.
Again, Theocritus may encounter his rustics
fluting in rivalry, like Daphnis and Menalcas
in the eighth **idyl**, 'on the **long** ranges of the
hills.'   Their **kine** and **sheep have** fed upwards
from the lower valleys to the place **where**

'The track **winds** down to the **clear stream,**
　To cross the sparkling shallows ; there
　The cattle love to gather, on their way
　To the high mountain-pastures, and to **stay,**
　Till the rough cow-herds drive them past,
　Knee-deep in the cool ford ; for 'tis the last
　Of all the woody, high, well-water'd dells
　On Etna,　　　.　　　.　　　.
　.　　　.　　　.　　　.　　　glade,
And stream, and **sward,** and chestnut-trees,
　End here ; Etna beyond, in the broad glare
　Of the hot noon, without a shade,
　Slope behind slope, up to the peak, lies bare ;
　The **peak,** round which the white clouds play.'[1]

Theocritus never drives his flock so high,

[1] *Empedocles on Etna.*

and rarely muses on such thoughts as come to
wanderers beyond the shade of trees and the
sound of water among the scorched rocks and
the barren lava.   The day is always cooled
and soothed, in his idyls, with the 'music of
water that falleth from the high face of the
rock,' or with the murmurs of the sea.   From
the cliffs and their seat among the bright red
berries on the arbutus shrubs, his shepherds
flute to each other, as they watch the tunny
fishers cruising far below, while the echo floats
upwards of the sailors' song.   These shepherds
have some touch in them of the satyr nature ;
we might fancy that their ears are pointed like
those of Hawthorne's Donatello, in ' Trans-
formation.'

It should be noticed, as a proof of the truth-
fulness of Theocritus, that the songs of his
shepherds and goatherds are all such as he
might really have heard on the shores of Sicily.
This is the real answer to the criticism which
calls him affected.   When mock pastorals
flourished at the court of France, when the
long dispute as to the merits of the ancients
and moderns was raging, critics vowed that the
hinds of Theocritus were too sentimental and
polite in their wooings.   Refinement and sentiment were to be reserved for princely shepherds
dancing, crook in hand, in the court ballets.
Louis XIV sang of himself—

> ' A son labeur il passe tout d'un coup,
> Et n'ira pas dormir sur la fougère,

Ny s'oublier aupres d'une Bergere,
Jusques au point d'en oublier le Loup.' [1]

Accustomed to royal goatherds in silk and
lace, Fontenelle (a severe critic of Theocritus)
could not believe in the delicacy of a Sicilian
who wore a skin 'stripped from the roughest
of he-goats, with the smell of the rennet clinging to it still.' Thus Fontenelle cries, 'Can
any one suppose that there ever was a shepherd who could say "Would I were the humming bee, Amaryllis, to flit to thy cave, and
dip beneath the branches, and the ivy leaves
that hide thee"?' and then he quotes other
graceful passages from the love-verses of Theocritean swains. Certainly no such fancies were
to be expected from the French peasants of
Fontenelle's age, 'creatures blackened with the
sun, and bowed with labour and hunger.' The
imaginative grace of Battus is quite as remote
from our own hinds. But we have the best reason to suppose that the peasants of Theocritus's
time expressed refined sentiment in language
adorned with colour and music, because the
modern love-songs of Greek shepherds sound like
memories of Theocritus. The lover of Amaryllis
might have sung this among his ditties—

Χελιδονάκι θὰ γενῶ, σ' τὰ χείλη σου νὰ καττῶ
Νὰ σὲ φιλήσω μιὰ καὶ δυό, καὶ πάλε νὰ πετάξω.

'To flit towards these lips of thine, I fain would be
  a swallow,

---

[1] Ballet des Arts, dansé par sa Majesté ; le 8 janvier,
1663. A Paris, par Robert Ballard. MDCLXIII.

To kiss thee once, to kiss thee twice, **and then** go
flying homeward.'[1]

In his despair, when Love 'clung **to him**
like a leech of the fen,' **he** might have **murmured**—

Ἤθελα νὰ εἶμαι σ' τὰ βουνά, μ'αλάφια νὰ κοιμοῦμαι
Καὶ τὸ δίκον σου τὸ κορμί νὰ μὴ τὸ συλλογιοῦμαι.

'Would that I were on the high hills, and lay
where lie the stags, and no more was troubled with
the thought of thee.'

Here, again, is a love-complaint from modern
Epirus, exactly in **the tone of Battus's** song in
the tenth idyl—

'White thou art not, thou art not golden haired,
  Thou **art** brown, and gracious, and meet for love.'

Here is a longer love-ditty—
'I will begin by telling thee first of thy per-
fections : thy body is as fair as an angel's ; no
painter could design it. And if any man be
sad, he has but to look on thee, and despite
himself he takes courage, the hapless one, and
his heart is joyous. Upon thy brows are shining
the constellated Pleiades, thy breast is full of
the flowers of May, thy breasts are lilies.
Thou hast the eyes of a princess, the glance of
a queen, and but one fault hast thou, that thou
deignest not **to speak to me.'**

---

[1] These and the following ditties are from the modern
Greek ballads collected by MM. Fauriel and Legrand.

Battus might have cried thus, with a modern Greek singer, to the shade of the dead Amaryllis (Idyl IV), the 'gracious Amaryllis, unforgotten even in death '—

'Ah, light of mine eyes, what gift shall I send thee ; what gift to the other world ? The apple rots, and the quince decayeth, and one by one **they perish,** the **petals of the rose ! I** send **thee my tears bound** in a **napkin, and what though** the napkin burns, if my tears reach thee at last !'

The difficulty is to stop choosing, where all the verses of the modern Greek peasants are so rich in Theocritean memories, so ardent, so delicate, so full of flowers and birds and the music of fountains. Enough has been said, perhaps, to show what the popular poetry of Sicily could lend to the genius of Theocritus.

From her shepherds he borrowed much,— their bucolic melody ; their love-complaints ; their rural superstitions; their system of answering couplets, in which each singer refines on the utterance of his rival. But he did not borrow their ' pastoral melancholy.' There is little of melancholy in Theocritus. When Battus **is chilled by the thought of** the death of Amaryllis, it is but as one is chilled when a thin cloud passes over the sun, on a bright day of early spring. And in an epigram the dead girl is spoken of as the kid that the wolf has seized, while the hounds bay all too late. Grief will not bring her back. The world must go its way, and we need not darken its sunlight by long regret. Yet when, for once, Theocritus adopted the accent of pastoral lament, when he raised the rural dirge for Daphnis into the realm of art, he composed a masterpiece, and a model for all later poets, as for the authors of *Lycidas*, *Thyrsis*, and *Adonais*.

Theocritus did more than borrow a note from the country people. He brought the gifts of his own spirit to the contemplation of the world. He had the clearest vision, and he had the most ardent love of poetry, 'of song may all my dwelling be full, for neither is sleep more sweet, nor sudden spring, nor are flowers more delicious to the bees, so dear to me are the Muses.' . . . 'Never may we be sundered, the Muses of Pieria and I.' Again, he had perhaps in greater measure than any other poet the gift of the undisturbed enjoyment of life. The undertone of all his idyls is joy in the sunshine and in existence. His favourite word, the word that opens the first idyl, and, as it were, strikes the keynote, is ἁδύ, *sweet*. He finds all things delectable in the rural life :

'Sweet are the voices of the calves, and sweet the heifers' lowing ; sweet plays the shepherd on the shepherd's pipe, and sweet is the echo.'

Even in courtly poems, and in the artificial hymns of which we are to speak in their place, the memory of the joyful country life comes over him. He praises Hiero, because Hiero is to restore peace to Syracuse, and when peace returns, then 'thousands of sheep fattened in the meadows will bleat along the plain, and the kine, as they flock in crowds to the stalls, will make the belated traveller hasten on his way.' The words evoke a memory of a narrow country lane in the summer evening, when light is dying out of the sky, and the fragrance of wild roses by the roadside is mingled with the perfumed breath of cattle that hurry past on their homeward road. There was scarcely a form of the life he saw that did not seem to him worthy of song, though it might be but the gossip of two rude hinds, or the drinking bout of the Thessalian horse-jobber, and the false girl Cynisca and her wild lover Æschines. But it is the sweet country that he loves best to behold and to remember. In his youth Sicily and Syracuse were disturbed by civil and foreign wars, wars of citizens against citizens, of Greeks against Carthaginians, and against the fierce 'men of Mars,' the banded mercenaries who possessed themselves of Messana. But this was not matter for his joyous Muse—

κεῖνος δ' οὐ πολέμους, οὐ δάκρυα, Πᾶνα δ' ἔμελπε,
καὶ βούτας ἐλίγαινε καὶ ἀείδων ἐνόμευε.

'Not of wars, not of tears, but of Pan would he chant, and of the neatherds he sweetly sang, and singing he shepherded his flocks.'

This was the training that Sicily, her hills, her seas, her lovers, her poet-shepherds, gave

to Theocritus. Sicily showed him subjects
which he imitated in truthful art. Unluckily
the later pastoral poets of northern lands have
imitated *him*, and so have gone far astray from
northern nature. The pupil of nature had still
to be taught the 'rules' of the critics, to watch
the temper and fashion of his time, and to try
his fortune among the courtly poets and grammarians of the capital of civilisation. Between
the years of early youth in Sicily and the years
of waiting for court patronage at Alexandria,
it seems probable that we must place a period
of education in the island of Cos. The testi-
monies of the Grammarians who handed on to
us the scanty traditions about Theocritus, agree
in making him the pupil of Philetas of Cos.
This Philetas was a critic, a commentator on
Homer, and an elegiac poet whose love-songs
were greatly admired by the Romans of the
Augustan age. He is said to have been the
tutor of Ptolemy Philadelphus, who was himself born, as Theocritus records, in the isle of
Cos. It has been conjectured that Ptolemy
and Theocritus were fellow pupils, and that the
poet may have hoped to obtain court favour at
Alexandria from this early connection. About
this point nothing is certainly known, nor can
we exactly understand the sort of education
that was given in the school of the poet
Philetas. The ideas of that artificial age make
it not improbable that Philetas professed to
teach the art of poetry. A French critic and
poet of our own time, M. Baudelaire, was willing

to do as much 'in thirty lessons.' Possibly
Philetas may have imparted technical rules
then in vogue, and the fashionable knack of
introducing obscure mythological allusions. He
was a logician as well as a poet, and is fabled
to have died of vexation because he could not
unriddle one of the metaphysical catches or
puzzles of the sophists. His varied activity
seems to have worn him to a shadow; the contemporary satirists bantered him about his lean-
ness, and it was alleged that he wore leaden
soles to his sandals lest the wind should blow
him, as it blew the calves of Daphnis (Idyl IX)
over a cliff against the rocks, or into the sea.[1]
Philetas seems a strange master for Theocritus,
but, whatever the qualities of the teacher, Cos,
the home of the luxurious old age of Meleager,
was a beautiful school. The island was one of
the most ancient colonies of the Dorians, and
the Syracusan scholar found himself among a
people who spoke his own broad and liquid
dialect. The sides of the limestone hills were
clothed with vines, and with shadowy plane-
trees which still attain extraordinary size and
age, while the wine-presses where Demeter
smiled, 'with sheaves and poppies in her hands,'
yielded a famous vintage. The people had a
soft industry of their own, they fashioned the
'Coan stuff,' transparent robes for woman's
wear, like the ὑδάτινα βράκη, the thin un-
dulating tissues which Theugenis was to weave

[1] See Couat, *La Poesie Alexandrine*, p. 68 *et seq.*,
Paris, 1882.

with the ivory distaff, the gift of Theocritus.
As a colony of Epidaurus, Cos naturally culti-
vated the worship of Asclepius, the divine
physician, the child of Apollo. In connection
with his worship and with the clan of the
Asclepiadae (that widespread stock to which
Aristotle belonged, and in which the practice
of leechcraft was hereditary), Cos possessed a
school of medicine. In the temple of Asclepius
patients hung up as votive offerings representations of their diseased limbs, and thus the
temple became a museum of anatomical speci-
mens. Cos was therefore resorted to by young
students from all **parts of the** East, and
Theocritus cannot but have made many friends
of his own age. Among these he alludes in
various passages to Nicias, afterwards a physician at Miletus, to Philinus, noted in later
life as the head of a medical sect, and to
Aratus. Theocritus has sung of Aratus's love-
affairs, and St. Paul has quoted him as a
witness to man's instinctive consent in the
doctrine of the universal fatherhood of God.
These strangely various notices have done
more for the memory of Aratus than his own
didactic poem on the meteorological theories
of his age. He lives, with Philinus and the
rest of the Coan students, because Theocritus
introduced them into the picture of a happy
summer's day. In the seventh idyl, that one
day of Demeter's harvest-feast is immortal, and
the sun never goes down on its delight. **We**
see Theocritus

κοὔπω τὰν μεσάταν ὁδὸν ἄνυμες, οὐδὲ τὸ σᾶμα
ἀμῖν τὸ Βρασίλα κατεφαίνετο—

when he 'had not yet reached the mid-point of
the way, nor had the tomb yet risen on his
sight.' He reveals himself as he was at the
height of morning, at the best moment of the
journey, in midsummer of a genius still un-
checked by doubt, or disappointment, or neglect.
Life seems to accost him with the glance of the
goatherd Lycidas, 'and still he smiled as he
spoke, with laughing eyes, and laughter dwelling on his lips.' In Cos, Theocritus found
friendship, and met Myrto, 'the girl he loved
as dearly as goats love the spring.' Here he
could express, without any afterthought, an enthusiastic adoration for the disinterested joys,
the enchanted moments of human existence.
Before he entered the thronged streets of Alexandria, and tuned his shepherd's pipe to catch
the ear of princes, and to sing the epithalamium
of a royal and incestuous love, he rested with
his friends in the happy island. Deep in a
cave, among the ruins of ancient aqueducts,
there still bubbles up, from the Coan limestone,
the wellspring of the Nymphs. 'There they
reclined on beds of fragrant rushes, lowly
strown, and rejoicing they lay in new stript
leaves of the vine. And high above their
heads waved many a poplar, many an elm-
tree, while close at hand the sacred water from
the nymph's own cave welled forth with murmurs musical' (Idyl VII).

The old Dorian settlers in Syracuse pleased
themselves with the fable that their fountain,
Arethusa, had been a Grecian nymph, who,
like themselves, had crossed the sea to Sicily.
The poetry of Theocritus, read or sung in
sultry Alexandria, must have seemed like a
new welling up of the waters of Arethusa in
the sandy soil of Egypt. We cannot certainly
say when the poet first came from Syracuse, or
from Cos, to Alexandria. It is evident however from the allusions in the fifteenth and
seventeenth idyls that he was living there after
Ptolemy Philadelphus married his own sister,
Arsinoë. It is not impossible to form some
idea of the condition of Alexandrian society,
art, religion, literature and learning at the court
of Ptolemy Philadelphus. The vast city, founded
some sixty years before, was now completed.
The walls, many miles in circuit, protected a
population of about eight hundred thousand
souls. Into that changing crowd were gathered
adventurers from all the known world. Mer-
chantmen brought to Ptolemy the wares of
India and the porcelains of China. Marauders
from upper Egypt skulked about the native
quarters, and sallied forth at night to rob the
wayfarer. The king's guards were recruited
with soldiers from turbulent Greece, from Asia,
from Italy. Settlers were attracted from Syracuse by the prospect of high wages and profit-
able labour. The Jewish quarters were full of
Israelites who did not disdain Greek learning.
The city in which this multitude found a home

was beautifully constructed. The Mediterranean filled the northern haven, the southern walls were washed by the Mareotic lake. If the isle of Pharos shone dazzling white, and wearied the eyes, there was shade beneath the long marble colonnades, and in the groves and cool halls of the Museum and the Libraries. The Etesian winds blew fresh in summer from the north, across the sea, and refreshed the people in their gardens. No town seemed greater nor wealthier to the voyager, who (like the hero of the Greek novel *Clitophon and Leucippe*) entered by the gate of the Sun, and found that, after nightfall, the torches borne by men and women hastening to some religious feast, filled the dusk with a light like that of 'the sun cut up into fragments.' At the same time no town was more in need of the memories of the country, which came to her in well-watered gardens, in landscape-paintings, and in the verse of Theocritus.

It is impossible to give a clearer idea of the opulence and luxury of Alexandria and her kings, than will be conveyed by the description of the coronation-feast of Ptolemy Philadelphus. This great masquerade and banquet was prepared by the elder Ptolemy on the occasion of his admitting his son to share his throne. The entertainment was described (in a work now lost) by Callixenus of Rhodes, and the record has been preserved by Atheneaus (v. 25). The inner pavilion in which the guests of Ptolemy reclined, contained one hundred and thirty-five

couches. Over the roof was placed a scarlet
awning, with a fringe of white, and there were
many other awnings, richly embroidered with
mythological designs. The pillars which sus-
tained the roof were **shaped** in the **likeness of
palm-trees, and of** *thyrsi*, the weapons **of the**
wine-god Dionysus. Round three outer **sides**
ran arcades, draped with purple tissues, **and**
with the skins of strange beasts. The fourth
side, open to the air, was shady with the foliage
of myrtles and laurels. Everywhere the ground
was carpeted with flowers, though the season
was mid-winter, with roses **and** white lilies and
blossoms of the gardens. By the columns
round the **whole** pavilion **were arrayed** a
hundred effigies in marble, **executed by the**
most famous sculptors, and on the middle
spaces were hung works by the painters **of**
Sicyon and tapestry woven with stories of **the**
adventures **of** the gods. Above these, again,
ran a frieze of gold and silver shields, while in
the higher niches were placed comic, tragic,
and satiric sculptured groups 'dressed in real
clothes,' says the historian, much admiring this
realism. It is impossible to number the tripods,
and flagons, and couches of gold, resting on
golden figures **of** sphinxes, the salvers, the
bowls, **the** jewelled vases. The masquerade
of this winter festival began with the procession
of **the** Morning - star, Heosphoros, **and** then
followed a masque of kings and **a revel of vari-**
ous gods, while the company of Hesperus, the
Evening-star followed, and ended all. The

revel of Dionysus was introduced by men dis-
guised as Sileni, wild woodland beings in
raiment of purple and scarlet.   Then came
scores of satyrs with gilded lamps in their
hands.   Next appeared beautiful maidens,
attired as Victories, waving golden wings and
swinging vessels of burning incense.   The
altar of the God of the Vine was borne behind
them, crowned and covered with leaves of
gold, and next boys in purple robes scattered
fragrant scents from golden salvers.   Then
came a throng of gold-crowned satyrs, their
naked bodies stained with purple and ver-
milion, and among them was a tall man who
represented the year and carried a horn of
plenty.   He was followed by a beautiful woman
in rich attire, carrying in one hand branches
of the palm-tree, in the other a rod of the
peach-tree, starred with its constellated flowers.
Then the masque of the Seasons swept by, and
Philiscus followed, Philiscus the Corcyraean,
the priest of Dionysus, and the favourite tragic
poet of the court.   After the prizes for the
athletes had been borne past, Dionysus himself was charioted along, a gigantic figure clad
in purple, and pouring libations out of a golden
goblet.   Around him lay huge drinking-cups,
and smoking censers of gold, and a bower of
vine leaves grew up, and shaded the head of
the god.   Then hurried by a crowd of priests
and priestesses, Maenads, Bacchantes, Bassa-
rids, women crowned with the vine, or with
garlands of snakes, and girls bearing the mystic

vannus Iacchi. And still the procession was
not ended. A mechanical figure of Nysa
passed, in a chariot drawn by eighty men,
among clusters of grapes formed of precious
stones, and the figure arose, and poured milk
out of a golden horn. The Satyrs and Sileni
followed close, and behind them six hundred
men dragged on a wain, a silver vessel that
held six hundred measures of wine. This was
only the first of countless symbolic vessels
that were carried past, till last came a multitude of sixteen hundred boys clad in white
tunics, and garlanded with ivy, who bore and
handed to the guests golden and silver vessels
full of sweet wine. All this was only part of
one procession, and the festival ended when
Ptolemy and Berenice and Ptolemy Philadelphus had been crowned with golden crowns
from many subject cities and lands.

This festival was obviously arranged to please
the taste of a prince with late Greek ideas of
pictorial display, and with barbaric wealth at
his command. Theocritus himself enables us
in the seventeenth idyl to estimate the opulence
and the dominion of Ptolemy. He was not
master of fertile Aegypt alone, where the Nile
breaks the rich dank soil, and where myriad
cities pour their taxes into his treasuries.
Ptolemy held lands also in Phoenicia, and
Arabia ; he claimed Syria and Libya and
Aethiopia ; he was lord of the distant Pamphylians, of the Cilicians, the Lycians and the
Carians, and the Cyclades owned his mastery.

Thus the wealth of the richest part of the
world flowed into Alexandria, attracting thither
the priests of strange religions, the possessors
of Greek learning, the painters and sculptors
whose work has left its traces on the genius of
Theocritus.

Looking at this early Alexandrian age, three
points become clear to us. First, the fashion
of the times was Oriental, Oriental in religion
and in society. Nothing could be less Hellenic
than the popular cult of Adonis. The fifteenth
idyl of Theocritus shows us Greek women wor-
shipping in their manner at an Assyrian shrine,
the shrine of that effeminate lover of Aphrodite,
whom Heracles, according to the Greek proverb,
thought 'no great divinity.' The hymn of Bion,
with its luxurious lament, was probably meant
to be chanted at just such a festival as Theocritus describes, while a crowd of foreigners
gossiped among the flowers and embroideries,
the strangely-shaped sacred cakes, the ebony, the
gold, and the ivory. Not so much Oriental as
barbarous was the impulse which made Ptolemy
Philadelphus choose his own sister, Arsinoë,
for wife, as if absolute dominion had already
filled the mind of the Macedonian royal race
with the incestuous pride of the Incas, or of
Queen Hatasu, in an elder Egyptian dynasty.
This nascent barbarism has touched a few of
the Alexandrian poems even of Theocritus, and
his panegyric of Ptolemy, of his divine ancestors,
and his sister-bride is not much more Greek in
sentiment than are those old native hymns of

Pentaur to 'the strong Bull,' or the 'Risen
Sun,' to Rameses or Thothmes.

Again, the early Alexandrian was what we
call a 'literary' age. Literature was not an
affair of religion and of the state, but ministered
to the pleasure of individuals, and at their
pleasure was composed.[1] The temper of the
time was crudely critical. The Museum and
the Libraries, with their hundreds of thousands
of volumes, were hot-houses of grammarians
and of learned poets. Callimachus, the head
librarian, was also the most eminent man of
letters. Unable, himself, to compose a poem
of epic length and copiousness, he discouraged
all long poems. He shone in epigrams, pedantic
hymns, and didactic verses. He toyed with
anagrams, and won court favour by discover-
ing that the letters of 'Arsinoë,' the name of
Ptolemy's wife, made the words ἴον Ἥρας, the
violet of Hera. In another masterpiece the
genius of Callimachus followed the stolen tress
of Queen Berenice to the skies, where the locks
became a constellation. A contemporary of
Callimachus was Zenodotus, the critic, who was
for improving the Iliad and Odyssey by cutting
out all the epic commonplaces which seemed
to him to be needless repetitions. It is pretty
plain that, in literary society, Homer was
thought out of date and *rococo*. The favourite
topics of poets were now, not the tales of Troy
and Thebes, but the amorous adventures of the
gods. When Apollonius Rhodius attempted to

---

[1] See Couat, *op. cit.* p. 395.

revive the epic, it is said that the influence of
Callimachus quite discomfited the young poet.
A war of epigrams began, and while Apollonius
called Callimachus a 'blockhead' (so finished
was his invective), the veteran compared his
rival to the Ibis, the scavenger-bird. Other
singers satirised each others' legs, and one, the
Aretino of the time, mocked at king Ptolemy
and scourged his failings in verse. The literary
quarrels (to which Theocritus seems to allude
in Idyl VII, where Lycidas says he 'hates the
birds of the Muses that cackle in vain rivalry
with Homer') were as stupid as such affairs
usually are. The taste for artificial epic was
to return; although many people already declared that Homer was the world's poet, and
that the world needed no other. This epic
reaction brought into favour Apollonius Rhodius,
author of the *Argonautica*. Theocritus has been
supposed to aim at him as a vain rival of Homer,
but M. Couat points out that Theocritus was
seventy when Apollonius began to write. The
literary fashions of Alexandria are only of
moment to us so far as they directly affected
Theocritus. They could not make him obscure,
affected, tedious, but his nature probably inclined him to obey fashion so far as only to
write short poems. His rural poems are εἰδύλ-
λια, 'little pictures.' His fragments of epic,
or imitations of the epic hymns are not

ὅσα πόντος ἀείδει

—not full and sonorous as the songs of Homer

and the sea. 'Ce poète est le moins naïf qui se puisse rencontrer, et il se dégage de son oeuvre un parfum de naïveté rustique.'[1] They are, what a German critic has called them, *mythologischen genre-bilder*, cabinet pictures in the manner called *genre*, full of pretty detail and domestic feeling. And this brings us to the third characteristic of the age, — its art was elaborately pictorial. Poetry seems to have sought inspiration from painting, while painting, as we have said, inclined to *genre*, to luxurious representations of the amours of the gods or the adventures of heroes, with backgrounds of pastoral landscape. Shepherds fluted while Perseus slew Medusa.

The old order of things in Greece had been precisely the opposite of this Alexandrian manner. Homer and the later Homeric legends, with the tragedians, inspired the sculptors, and even the artisans who decorated vases. When a new order of subjects became fashionable, and when every rich Alexandrian had pictures or frescoes on his walls, it appears that the painters took the lead, that the initiative in art was theirs. The Alexandrian pictures perished long ago, but the relics of Alexandrian style which remain in the buried cities of Campania, in Pompeii especially, bear testimony to the taste of the period.[2] Out of nearly two thousand Pompeian pictures, it is

[1] Couat, p. 434.
[2] See Helbig, *Campanische Wandmalerei*, and Brunn, *Die griechischen Bukoliker und die Bildende Kunst.*

calculated that some fourteen hundred (roughly speaking) are mythological in subject. The loves of the gods are repeated in scores of designs, and these designs closely correspond to the mythological poems of Theocritus and his younger contemporaries Bion and Moschus. Take as an example the adventure of Europa : Lord Tennyson's lines, in *The Palace of Art* are intended to describe a *picture—*

'Or sweet Europa's mantle blew unclasp'd,
  From off her shoulder backward borne :
From one hand droop'd a crocus : one hand grasp'd
    The mild bull's golden horn.'

The words of Moschus also seem as if they might have derived their inspiration from a painting, the touches are so minute, and so picturesque—

'Meanwhile Europa, riding on the back of the divine bull, with one hand clasped the beast's great horn, and with the other caught up her garment's purple fold, lest it might trail and be drenched in the hoar sea's infinite spray. And her deep robe was blown out in the wind, like the sail of a ship, and lightly ever it wafted the maiden onward.'

Now every single 'motive' of this description, —Europa with one hand holding the bull's horn, with the other lifting her dress, the wind puffing out her shawl like a sail, is repeated in the Pompeian wall-pictures, which themselves are believed to be derived from Alexandrian originals. There are more curious coincidences

than this.   In the sixth idyl of Theocritus,
Damoetas makes the Cyclops say that Galatea
'will send him many a messenger.'   The mere
idea of describing the monstrous cannibal Polyphemus in love, is artificial and Alexandrian.
But who were the 'messengers' of the sea-
nymph Galatea?   A Pompeian picture illustrates
the point, by representing a little Love riding
up to the shore on the back of a dolphin, with
a letter in his hand for Polyphemus.   Greek
art in Egypt suffered from an Egyptian plague
of Loves.   Loves flutter through the Pompeian
pictures as they do through the poems of Moschus and Bion.   They are carried about in
cages, for sale, like birds.   They are caught in
bird-traps.   They don the lion-skin of Heracles.
They flutter about baskets laden with roses ;
round rosy Loves, like the cupids of Boucher.
They are not akin to 'the grievous Love,' the
mighty wrestler who threw Daphnis a fall, in
the first idyl of Theocritus.   They are 'the
children that flit overhead, the little Loves,
like the young nightingales upon the budding
trees,' which flit round the dead Adonis in the
fifteenth idyl.   They are the birds that shun
the boy fowler, in Bion's poem, and perch
uncalled (as in a bronze in the Uffizi) on the
grown man.   In one or other of the sixteen
Pompeian pictures of Venus and Adonis, the
Loves are breaking their bows and arrows for
grief, as in the hymn of Bion.

Enough has perhaps been said about the
social and artistic taste of Alexandria to account

for the remarkable differences in manner between the rustic idyls of Theocritus and the
epic idyls of himself and his followers Moschus
and Bion. In the rural idyls, Theocritus was
himself, and wrote to please himself. In the
epic idyls, as in the Hymn to the Dioscuri,
and in the two poems on Heracles, he was
writing to please the taste of Alexandria. He
had to choose epic topics, but he was warned
by the famous saying of Callimachus ('a great
book is a great evil') not to imitate the length
of the epic.¹ He was also to shun close imita-
tion of what are so easily imitated, the regular
recurring *formulae*, the commonplace of Homer.
He was to add minute pictorial touches, as in
the description of Alcmena's waking when the
serpents attacked her child,—a passage rich in
domestic pathos and incident which contrast
strongly with Pindar's bare narrative of the
same events. We have noted the same pictorial
quality in the *Europa* of Moschus. Our own
age has often been compared to the Alexandrian
epoch, to that era of large cities, wealth, refinement, criticism, and science ; and the pictorial
Idylls of the King very closely resemble the
epico-idyllic manner of Alexandria. We have
tried to examine the society in which Theocritus
lived. But our impressions about the poet are
more distinct. In him we find the most genial
character ; pious as Greece counted piety ;

¹ The *Hecale* of Callimachus, or Theseus and the
Marathonian Bull, seems to have been rather a heroic
idyl than an epic.

tender as became the poet of love ; glad as
the singer of a happy southern world should
be ; gifted, above all, with humour, and with
dramatic power. 'His lyre has all the chords';
his is the last of all the perfect voices of Hellas ;
after him no man saw life with eyes so steady
and so mirthful.

About the lives of the three idyllic poets
literary history says little. About their deaths
she only tells us through the dirge by Moschus,
that Bion was poisoned. The lovers of Theocritus would willingly hope that he returned
from Alexandria to Sicily, about the time when
he wrote the sixteenth idyl, and that he lived
in the enjoyment of the friendship and the
domestic happiness and honour which he sang
so well, through the golden age of Hiero (264
B.C.) No happier fortune could befall him
who wrote the epigram of the lady of heavenly
love, who worshipped with the noble wife of
Nicias under the green roof of Milesian Aphrodite, and who prophesied of the return of peace
and of song to Sicily and Syracuse.

THEOCRITUS

B

THEOCRITUS

IDYL I

The shepherd Thyrsis meets a goatherd, in a shady place beside a spring, and at his invitation sings the Song of Daphnis. *This ideal hero of Greek pastoral song had won for his bride the fairest of the Nymphs. Confident in the strength of his passion, he boasted that Love could never subdue him to a new affection. Love avenged himself by making Daphnis desire a strange maiden, but to this temptation he never yielded, and so died a constant lover.* The song *tells how the cattle and the wild things of the wood bewailed him, how Hermes and Priapus gave him counsel in vain, and how with his last breath he retorted the taunts of the implacable Aphrodite.* The scene *is in Sicily.*

Thyrsis. Sweet, meseems, is the whispering sound of yonder pine tree, goatherd, that murmureth by the wells of water; and sweet are thy pipings. After Pan the second prize shalt thou bear away, and if he take the hornéd goat, the she-goat shalt thou win; but if he choose the she-goat for his meed, the kid

falls to thee, and dainty is the flesh of kids
e'er the age when thou milkest them.

The Goatherd. Sweeter, O shepherd, is thy
song than the music of yonder water that is
poured from the high face of the rock! Yea,
if the Muses take the young ewe for their
gift, a stall-fed lamb shalt thou receive for thy
meed; but if it please them to take the lamb,
thou shalt lead away the ewe for the second
prize.

Thyrsis. Wilt thou, goatherd, in the nymphs'
name, wilt thou sit thee down here, among
the tamarisks, on this sloping knoll, and pipe
while in this place I watch thy flocks?

Goatherd. Nay, shepherd, it may not be;
we may not pipe in the noontide. 'Tis Pan
we dread, who truly at this hour rests weary
from the chase; and bitter of mood is he,
the keen wrath sitting ever at his nostrils.
But, Thyrsis, for that thou surely wert wont
to sing *The Affliction of Daphnis,* and hast
most deeply meditated the pastoral muse,
come hither, and beneath yonder elm let us
sit down, in face of Priapus and the fountain
fairies, where is that resting-place of the
shepherds, and where the oak trees are. Ah!
if thou wilt but sing as on that day thou
sangest in thy match with Chromis out of
Libya, I will let thee milk, ay, three times, a
goat that is the mother of twins, and even
when she has suckled her kids her milk doth
fill two pails. A deep bowl of ivy-wood, too,
I will give thee, rubbed with sweet bees'-wax, a

twy-eared bowl newly wrought, smacking still
of the knife of the graver. Round its upper
edges goes the ivy winding, ivy besprent with
golden flowers ; and about it is a tendril twisted
that joys in its saffron fruit. Within Is designed a maiden, as fair a thing as the gods
could fashion, arrayed in a sweeping robe, and
a snood on her head. Beside her two youths
with fair love-locks are contending from either
side, with alternate speech, but her heart
thereby is all untouched. And now on one
she glances, smiling, and anon she lightly
flings the other a thought, while by reason of
the long vigils of love their eyes are heavy, but
their labour is all in vain.

Beyond these an ancient fisherman and a
rock are fashioned, a rugged rock, whereon
with might and main the old man drags a
great net for his cast, as one that labours
stoutly. Thou wouldst say that he is fishing
with all the might of his limbs, so big the
sinews swell all about his neck, grey-haired
though he be, but his strength is as the strength
of youth. Now divided but a little space from
the sea-worn old man is a vineyard laden well with
fire-red clusters, and on the rough wall a little
lad watches the vineyard, sitting there. Round
him two she-foxes are skulking, and one goes
along the vine-rows to devour the ripe grapes,
and the other brings all her cunning to bear
against the scrip, and vows she will never
leave the lad, till she strand him bare and
breakfastless. But the boy is plaiting a pretty

locust-cage with stalks of asphodel, and fitting
it with reeds, and less care of his scrip has
he, and of the vines, than **delight** in his
plaiting.

All about the cup is spread the **soft acanthus**,
a miracle **of** varied work,[1] a thing for thee to
marvel on. For this bowl I paid to a Caly-
donian ferryman a goat and a great white
cream cheese. Never has its lip touched mine,
but it still lies maiden for me. Gladly with
this cup would I gain thee to my desire, if thou,
my friend, wilt sing **me that** delightful song.
Nay, I grudge it **thee not at all.** Begin, my
friend, for be sure **thou canst in no wise** carry
thy **song with thee to Hades, that** puts all
things out of mind !

The Song of Thyrsis.

Begin, ye Muses dear, begin the pastoral song!
Thyrsis of Etna am I, and this is the voice of
Thyrsis. Where, ah ! where were ye when
Daphnis was languishing ; ye Nymphs, where
were ye ? By Peneus's beautiful dells, or by
dells of Pindus ? for surely ye dwelt not by the
great stream of the river Anapus, nor on the
watch-tower of Etna, nor by the sacred water
of Acis.

*Begin, ye **Muses dear, begin the pastoral song!***

For him the jackals, for him the wolves did
cry ; for him did even the lion out of the forest

---

[1] Or, reading Αἰολικὸν = Aeolian, cf. Thucyd. iii. 102.

lament. Kine and bulls by his feet right many, and heifers plenty, with the young calves bewailed him.

Begin, ye Muses dear, begin the pastoral song!

Came Hermes first from the hill, and said, 'Daphnis, who is it that torments thee; child, whom dost thou love with so great desire?' The neatherds came, and the shepherds; the goatherds came: all they asked what ailed him. Came also Priapus,—

Begin, ye Muses dear, begin the pastoral song!

And said: 'Unhappy Daphnis, wherefore dost thou languish, while for thee the maiden by all the fountains, through all the glades is fleeting, in search of thee? Ah! thou art too laggard a lover, and thou nothing availest! A neatherd wert thou named, and now thou art like the goatherd:

Begin, ye Muses dear, begin the pastoral song!

'For the goatherd, when he marks the young goats at their pastime, looks on with yearning eyes, and fain would be even as they; and thou, when thou beholdest the laughter of maidens, dost gaze with yearning eyes, for that thou dost not join their dances.'

Begin, ye Muses dear, begin the pastoral song!

Yet these the herdsman answered not again, but he bare his bitter love to the end, yea, to the fated end he bare it.

Begin, ye Muses dear, begin the pastoral song!

Ay, but she too came, the sweetly smiling Cypris, craftily smiling she came, yet keeping her heavy anger; and **she** spake, saying : 'Daphnis, methinks thou didst boast **that** thou wouldst throw Love a fall, nay, is it not thyself that hast been thrown by grievous Love ?'

Begin, ye Muses dear, begin the pastoral song!

But to her Daphnis answered again : 'Implacable Cypris, Cypris terrible, Cypris of mortals detested, already dost thou deem that my latest sun has set; nay, Daphnis even in Hades shall prove great sorrow to Love.

Begin, ye Muses dear, begin the pastoral song!

'Where it is told how the herdsman with Cypris—— Get thee to Ida, get thee to Anchises ! There are oak trees—here only galingale blows, here sweetly hum the bees about the hives !

Begin, ye Muses dear, begin the pastoral song!

' Thine Adonis, too, is in his bloom, for he herds the sheep and slays the hares, and he chases all the wild beasts. Nay, go and confront Diomedes again, and say, " The herdsman Daphnis I conquered, do thou join battle with me."

Begin, ye Muses dear, begin the pastoral song!

'Ye wolves, ye jackals, and ye bears in the mountain caves, farewell ! The herdsman Daphnis ye never shall see again, no more in the dells, no more in the groves, no more in the woodlands. Farewell Arethusa, ye rivers, good-night, that pour down Thymbris your beautiful waters.

Begin, ye Muses dear, begin the pastoral song!

'That Daphnis am I who here do herd the kine, Daphnis who water here the bulls and calves.

'O Pan, Pan! whether thou art on the high hills of Lycaeus, or rangest mighty Maenalus, haste hither to the Sicilian isle! Leave the tomb of Helice, leave that high cairn of the son of Lycaon, which seems wondrous fair, even in the eyes of the blessed.[1]

Give o'er, ye Muses, come, give o'er the pastoral song!

'Come hither, my prince, and take this fair pipe, honey-breathed with wax-stopped joints; and well it fits thy lip: for verily I, even I, by Love am now haled to Hades.

Give o'er, ye Muses, come, give o'er the pastoral song!

'Now violets bear, ye brambles, ye thorns bear violets; and let fair narcissus bloom on the boughs of juniper! Let all things with all be confounded,—from pines let men gather pears, for Daphnis is dying! Let the stag

[1] These are places famous in the oldest legends of Arcadia.

drag down the hounds, let owls from the hills contend in song with the nightingales.'

Give o'er, ye Muses, come, give o'er the pastoral song!

So Daphnis spake, and ended; but fain would Aphrodite have given him back to life. Nay, spun was all the thread that the Fates assigned, and Daphnis went down the stream. The whirling wave closed over the man the Muses loved, the man not hated of the nymphs.

Give o'er, ye Muses, come, give o'er the pastoral song!

And thou, give me the bowl, and the she-goat, that I may milk her and pour forth a libation to the Muses. Farewell, oh, farewells manifold, ye Muses, and I, some future day, will sing you yet a sweeter song.

The Goatherd. Filled may thy fair mouth be with honey, Thyrsis, and filled with the honey-comb; and the sweet dried fig mayst thou eat of Aegilus, for thou vanquishest the cicala in song! Lo here is thy cup, see, my friend, of how pleasant a savour! Thou wilt think it has been dipped in the well-spring of the Hours. Hither, hither, Cissaetha: do thou milk her, Thyrsis. And you young she-goats, wanton not so wildly lest you bring up the he-goat against you.

IDYL II

Simaetha, madly in love with Delphis, who has for-
saken her, endeavours to subdue him to her by
magic, and by invoking the Moon, in her character
of Hecate, and of Selene. She tells the tale of the
growth of her passion, and vows vengeance if her
magic arts are unsuccessful.
The scene is probably some garden beneath the moonlit
sky, near the town, and within sound of the sea.
The characters are Simaetha, and Thestylis, her
handmaid.

WHERE are my laurel leaves? come, bring
them, Thestylis; and where are the love-
charms? Wreath the bowl with bright-red
wool, that I may knit the witch-knots against
my grievous lover,[1] who for twelve days, oh
cruel, has never come hither, nor knows
whether I am alive or dead, nor has once
knocked at my door, unkind that he is! Hath
Love flown off with his light desires by some
other path—Love and Aphrodite? To-morrow I will go to the wrestling school of Timagetus, to see my love and to reproach him with
all the wrong he is doing me. But now I will

---

[1] Reading καταδήσομαι. Cf. Fritzsche's note, and
Harpocration, s.v.

bewitch him with my enchantments ! Do thou, Selene, shine clear and fair, for softly, Goddess, to thee will I sing, and to Hecate of hell. The very whelps shiver before her as she fares through black blood and across the barrows of the dead.

Hail, awful Hecate ! to the end be thou of our company, and make this medicine of mine no weaker than the spells of Circe, or of Medea, or of Perimede of the golden hair.

My magic wheel, draw home to me the man I love !

Lo, how the barley grain first smoulders in the fire,—nay, toss on the barley, Thestylis ! Miserable maid, where are thy wits wandering ? Even to thee, wretched that I am, have I become a laughing-stock, even to thee ? Scatter the grain, and cry thus the while, ''Tis the bones of Delphis I am scattering !'

My magic wheel, draw home to me the man I love !

Delphis troubled me, and I against Delphis am burning this laurel ; and even as it crackles loudly when it has caught the flame, and suddenly is burned up, and we see not even the dust thereof, lo, even thus may the flesh of Delphis waste in the burning !

My magic wheel, draw home to me the man I love !

Even as I melt this wax, with the god to aid, so speedily may he by love be molten, the

Myndian Delphis! And as whirls this brazen wheel,[1] so restless, under Aphrodite's spell, may he turn and turn about my doors.

My magic wheel, draw home to me the man I love!

Now will I burn the husks, and thou, O Artemis, hast power to move hell's adamantine gates, and all else that is as stubborn. Thestylis, hark, 'tis so; the hounds are baying up and down the town! The Goddess stands where the three ways meet! Hasten, and clash the brazen cymbals.

My magic wheel, draw home to me the man I love!

Lo, silent is the deep, and silent the winds, but never silent the torment in my breast. Nay, I am all on fire for him that made me, miserable me, no wife but a shameful thing, a girl no more a maiden.

My magic wheel, draw home to me the man I love!

Three times do I pour libation, and thrice, my Lady Moon, I speak this spell:—Be it with a friend that he lingers, be it with a leman he lies, may he as clean forget them as Theseus, of old, in Dia—so legends tell—did utterly forget the fair-tressed Ariadne.

My magic wheel, draw home to me the man I love!

---

[1] On the word ῥόμβος, see Lobeck, *Aglaoph.* p. 700; and 'The Bull Roarer,' in the translator's *Custom and Myth.*

Coltsfoot is an Arcadian weed that maddens, on the hills, the young stallions and fleet-footed mares. Ah! even as these may I see Delphis; and to this house of mine, may he speed like a madman, leaving the bright palaestra.

My magic wheel, draw home to me the man I love!

This fringe from his cloak Delphis lost; that now I shred and cast into the cruel flame. Ah, ah, thou torturing Love, why clingest thou to me like a leech of the fen, and drainest all the black blood from my body?

My magic wheel, draw home to me the man I love!

Lo, I will crush an eft, and a venomous draught to-morrow I will bring thee!

But now, Thestylis, take these magic herbs and secretly smear the juice on the jambs of his gate (whereat, even now, my heart is captive, though nothing he recks of me), and spit and whisper, ' 'Tis the bones of Delphis that I smear.'

My magic wheel, draw home to me the man I love!

And now that I am alone, whence shall I begin to bewail my love? Whence shall I take up the tale: who brought on me this sorrow? The maiden-bearer of the mystic vessel came our way, Anaxo, daughter of Eubulus, to the grove of Artemis; and behold, she had many other wild beasts paraded for that

time, in the sacred show, and among them a
lioness.

*Bethink thee of my love, and whence it came,
my Lady Moon!*

And the Thracian servant of Theucharidas,
—my nurse that is but lately dead, and who
then dwelt at our doors,—besought me and
implored me to come and see the show. And I
went with her, wretched woman that I am, clad
about in a fair and sweeping linen stole, over
which I had thrown the holiday dress of Clearista.

*Bethink thee of my love, and whence it came,
my Lady Moon!*

Lo! I was now come to the mid-point of
the highway, near the dwelling of Lycon, and
there I saw Delphis and Eudamippus walking
together. Their beards were more golden than
the golden flower of the ivy; their breasts
(they coming fresh from the glorious wrestler's
toil) were brighter of sheen than thyself,
Selene!

*Bethink thee of my love, and whence it came,
my Lady Moon!*

Even as I looked I loved, loved madly, and
all my heart was wounded, woe is me, and my
beauty began to wane. No more heed took
I of that show, and how I came home I know
not; but some parching fever utterly overthrew
me, and I lay a-bed ten days and ten nights.

*Bethink thee of my love, and whence it came,
my Lady Moon!*

And oftentimes my skin waxed wan as the colour of boxwood, and all my hair was falling from my head, and what was left of me was but skin and bones. Was there a wizard to whom I did not seek, or a crone to whose house I did not resort, of them that have art magical? But this was no light malady, and the time went fleeting on.

Bethink thee of my love, and whence it came, my Lady Moon!

Thus I told the true story to my maiden, and said, 'Go, Thestylis, and find me some remedy for this sore disease. Ah me, the Myndian possesses me, body and soul! Nay, depart, and watch by the wrestling-ground of Timagetus, for there is his resort, and there he loves to loiter.

Bethink thee of my love, and whence it came, my Lady Moon!

'And when thou art sure he is alone, nod to him secretly, and say, "Simaetha bids thee to come to her," and lead him hither privily.' So I spoke; and she went and brought the bright-limbed Delphis to my house. But I, when I beheld him just crossing the threshold of the door, with his light step,—

Bethink thee of my love, and whence it came, my Lady Moon!

Grew colder all than snow, and the sweat streamed from my brow like the dank dews, and I had no strength to speak, nay, nor to

utter as much as children murmur in their
slumber, calling to their mother dear: and all
my fair body turned stiff as a puppet of wax.

*Bethink thee of my love, and whence it came,
my Lady Moon!*

Then when he had gazed **on me, he that**
knows not love, he fixed **his** eyes on **the**
ground, and **sat** down **on my** bed, and **spake
as he sat him down** : ' Truly, Simaetha, **thou**
didst by no more outrun mine own coming
hither, when thou badst me to thy roof, than of
late I outran in the race the beautiful Philinus:

*Bethink thee of my love, and whence it came,
my Lady Moon!*

' For I should have come ; yea, by sweet
Love, I should have come, with friends of mine,
two or three, as soon as night drew on, bearing
in my breast the apples of Dionysus, and on
my head silvery poplar leaves, the holy boughs
of Heracles, all twined with bands of purple.

*Bethink thee of my love, and whence it came,
my Lady Moon!*

' And **if you** had received me, they would
have taken **it** well, for among all the youths un-
wed I have a name for beauty and speed of foot.
With one kiss of thy lovely mouth I **had** been
content; but an if ye had thrust me forth, and the
door had been fastened with the bar, then truly
should torch and axe have broken in upon you.

*Bethink thee of my love, and whence it came,
my Lady Moon!*

C

'And now to Cypris first, methinks, my
thanks are due, and after Cypris it is thou that
hast caught me, lady, from the burning, in that
thou badst me come to this thy house, half consumed as I am! Yea, Love, 'tis plain, lights
oft a fiercer blaze than Hephaestus the God
of Lipara.

Bethink thee of my love, and whence it came,
my Lady Moon!

'With his madness dire, he scares both the
maiden from her bower and the bride from
the bridal bed, yet warm with the body of her
lord!'

So he spake, and I, that was easy to win,
took his hand, and drew him down on the soft
bed beside me. And immediately body from
body caught fire, and our faces glowed as they
had not done, and sweetly we murmured. And
now, dear Selene, to tell thee no long tale, the
great rites were accomplished, and we twain
came to our desire. Faultless was I in his
sight, till yesterday, and he, again, in mine.
But there came to me the mother of Philista,
my flute player, and the mother of Melixo,
to-day, when the horses of the Sun were climb-
ing the sky, bearing Dawn of the rosy arms
from the ocean stream. Many another thing
she told me; and chiefly this, that Delphis is a
lover, and whom he loves she vowed she knew
not surely, but this only, that ever he filled up
his cup with the unmixed wine, to drink a toast
to his dearest. And at last he went off hastily,

saying that he would cover with garlands the dwelling of his love.

This news my visitor told me, and she speaks the truth. For indeed, at other seasons, he would come to me thrice, or four times, in the day, and often would leave with me his Dorian oil flask. But now it is the twelfth day since I have even looked on him ! Can it be that he has not some other delight, and has forgotten me ? Now with magic rites I will strive to bind him,[1] but if still he vexes me, he shall beat, by the Fates I vow it, at the gate of Hell. Such evil medicines I store against him in a certain coffer, the use whereof, my lady, an Assyrian stranger taught me.

But do thou farewell, and turn thy steeds to Ocean, Lady, and my pain I will bear, as even till now I have endured it. Farewell, Selene bright and fair, farewell ye other stars, that follow the wheels of quiet Night.

---

[1] Reading καταδήσομαι. Cf. line 3, and note.

IDYL III

*A goatherd, leaving his goats to feed on the hillside, in
the charge of Tityrus, approaches the cavern of
Amaryllis, with its veil of ferns and ivy, and at-
tempts to win back the heart of the girl by song.
He mingles promises with harmless threats, and re-
peats, in exquisite verses, the names of the famous
lovers of old days, Milanion and Endymion. Fail-
ing to move Amaryllis, the goatherd threatens to
die where he has thrown himself down, beneath the
trees.*

COURTING Amaryllis with song I go, while my
she-goats feed on the hill, and Tityrus herds
them. Ah, Tityrus, my dearly beloved, feed
thou the goats, and to the well-side lead them,
Tityrus, and 'ware the yellow Libyan he-goat,
lest he butt thee with his horns.

Ah, lovely Amaryllis, why no more, as of
old, dost thou glance through this cavern after
me, nor callest me, thy sweetheart, to thy side.
Can it be that thou hatest me? Do I seem
snub-nosed, now thou hast seen me near,
maiden, and under-hung? Thou wilt make me
strangle myself!

Lo, ten apples I bring thee, plucked from
that very place where thou didst bid me

pluck them, and others to-morrow I will bring thee.

Ah, regard my heart's deep sorrow! ah, would I were that humming bee, and to **thy** cave might come dipping beneath the fern that hides thee, and the ivy leaves!

Now **know I** Love, **and a cruel God is he.** Surely **he sucked** the lioness's dug, and **in the** wild **wood** his mother **reared** him, whose fire is scorching me, and bites even to the bone.

Ah, lovely as thou art to look upon, ah heart of stone, ah dark-browed maiden, embrace me, thy true goatherd, that I may kiss thee, and even in empty kisses there is a sweet delight!

Soon wilt thou make me rend the wreath in pieces small, the wreath of ivy, dear Amaryllis, that I keep for thee, with rose-buds twined, and fragrant parsley. Ah me, what anguish! Wretched that I am, whither shall **I turn!** Thou dost not hear my prayer!

I will cast off my coat of skins, and into yonder waves I will spring, where the fisher Olpis watches for the tunny shoals, and even if I die not, surely thy pleasure will have been done.

I learned the truth of old, when, amid thoughts of thee, I asked, 'Loves she, loves she not?' and the poppy petal clung not, and gave **no** crackling sound, but withered on my smooth forearm, even so.[1]

And she too spoke sooth, even Agroeo, she that divineth with a sieve, and of late was binding sheaves behind the reapers, who said that

[1] He refers to a piece of folk-lore.

I had set all my heart on thee, but that thou didst nothing regard me.

Truly I keep for thee the white goat with the twin kids that Mermnon's daughter too, the brown-skinned Erithacis, prays me to give her ; and **give her them** I will, since thou dost flout me.

My right eyelid throbs, is it a sign that I am to see her ? Here will I lean me against this pine tree, and sing, and then perchance she will regard me, for she is not all of adamant.

Lo, Hippomenes when he was eager to marry the famous maiden, took apples in his hand, **and** so accomplished **his** course ; and Atalanta saw, and madly **longed,** and leaped into the deep waters of desire. Melampus too, the soothsayer, brought the herd of oxen from Othrys to Pylos, and thus in the arms of Bias was laid the lovely mother of wise Alphesiboea.

And was it not thus that Adonis, as he pastured his sheep upon the hills, led beautiful Cytherea to such heights of frenzy, that not even in his death doth she unclasp him from her bosom ? Blessed, methinks is the lot of him that sleeps, and tosses not, nor turns, even Endymion ; and, dearest maiden, blessed I call Iason, whom such things befell, as ye that be profane shall never come to know.

My head aches, but thou carest not. I will sing no more, but dead will I lie where I fall, and here may the wolves devour me.

Sweet as honey in the mouth may my death be to thee.

IDYL IV

Battus and Corydon, two rustic fellows, meeting in a glade, gossip about their neighbour, Aegon, who has gone to try his fortune at the Olympic games. After some random banter, the talk turns on the death of Amaryllis, and the grief of Battus is disturbed by the roaming of his cattle. Corydon removes a thorn that has run into his friend's foot, and the conversation comes back to matters of rural scandal.
The scene is in Southern Italy.

Battus. Tell me, Corydon, whose kine are these,—the cattle of Philondas ?

Corydon. Nay, they are Aegon's, he gave me them to pasture.

Battus. Dost thou ever find a way to milk them all, on the sly, just before evening ?

Corydon. No chance of that, for the old man puts the calves beneath their dams, and keeps watch on me.

Battus. But the neatherd himself,—to what land has he passed out of sight ?

Corydon. Hast thou not heard ? Milon went and carried him off to the Alpheus.

Battus. And when, pray, did *he* ever set eyes on the wrestlers' oil ?

Corydon. They say he is a match for Heracles, in strength and hardihood.

Battus. And I, so mother says, am a better man than Polydeuces.

Corydon. Well, off he has gone, with a shovel, and with twenty sheep from his flock here.[1]

Battus. Milo, thou'lt see, will soon be coaxing the wolves to rave!

Corydon. But Aegon's heifers here are lowing pitifully, and miss their master.

Battus. Yes, wretched beasts that they are, how false a neatherd was theirs!

Corydon. Wretched enough in truth, and they have no more care to pasture.

Battus. Nothing is left, now, of that heifer, look you, bones, that's all. She does not live on dewdrops, does she, like the grasshopper?

Corydon. No, by Earth, for sometimes I take her to graze by the banks of Aesarus, fair handfuls of fresh grass I give her too, and otherwhiles she wantons in the deep shade round Latymnus.

Battus. How lean is the red bull too! May the sons of Lampriades, the burghers to wit, get such another for their sacrifice to Hera, for the township is an ill neighbour.

Corydon. And yet that bull is driven to the mere's mouth, and to the meadows of Physcus, and to the Neaethus, where all fair herbs bloom, red goat-wort, and endive, and fragrant bees-wort.

---

[1] The shovel was used for tossing the sand of the lists; the sheep were food for Aegon's great appetite.

Battus. Ah, wretched Aegon, thy very kine will go to Hades, while thou too art in love with a luckless victory, and thy pipe is flecked with mildew, the pipe that once thou madest for thyself!

Corydon. Not the pipe, by the nymphs, not so, for when he went to Pisa, he left the same as a gift to me, and I am something of a player. Well can I strike up the air of *Glaucé*, and well the strain of *Pyrrhus*, and *the praise of Croton I sing*, and *Zacynthus is a goodly town*, and *Lacinium that fronts the dawn!* There Aegon the boxer, unaided, devoured eighty cakes to his own share, and there he caught the bull by the hoof, and brought him from the mountain, and gave him to Amaryllis. Thereon the women shrieked aloud, and the neatherd,—he burst out laughing.

Battus. Ah, gracious Amaryllis! Thee alone even in death will we ne'er forget. Dear to me as my goats wert thou, and thou art dead! Alas, too cruel a spirit hath my lot in his keeping.

Corydon. Dear Battus, thou must needs be comforted. The morrow perchance will bring better fortune. The living may hope, the dead alone are hopeless. Zeus now shows bright and clear, and anon he rains.

Battus. Enough of thy comforting! Drive the calves from the lower ground, the cursed beasts are grazing on the olive-shoots. Hie on, white face.

Corydon. Out, Cymaetha, get thee to the

hill! Dost thou not hear? By Pan, I will
soon come and be the death of you, if you stay
there! Look, here she is creeping back again!
Would I had my crook for hare killing: how
I would cudgel thee.

Battus. In the name of Zeus, prithee look
here, Corydon! A thorn has just run into my
foot under the ankle. How deep they grow,
the arrow-headed thorns. An ill end befall the
heifer; I was pricked when I was gaping after
her. Prithee dost see it?

Corydon. Yes, yes, and I have caught it in
my nails, see, here it is.

Battus. How tiny is the wound, and how
tall a man it masters!

Corydon. When thou goest to the hill, go
not barefoot, Battus, for on the hillside flourish
thorns and brambles plenty.

Battus. Come, tell me, Corydon, the old
man now, does he still run after that little
black-browed darling whom he used to dote
on?

Corydon. He is after her still, my lad; but
yesterday I came upon them, by the very byre,
and right loving were they.

Battus. Well done, thou ancient lover! Sure,
thou art near akin to the satyrs, or a rival of
the slim-shanked Pans![1]

---

[1] Reading ἐρίσδεις.

IDYL V

*This Idyl begins with a ribald debate between two hire-
lings, who, at last, compete with each other in a
match of pastoral song. No other idyl of Theocritus is so frankly true to the rough side of rustic
manners. The scene is in Southern Italy.*

Comatas. Goats of mine, keep clear of that
notorious shepherd of Sibyrtas, that Lacon ; he
stole my goat-skin yesterday.

Lacon. Will ye never leave the well-head ?
Off, my lambs, see ye not Comatas ; him that
lately stole my shepherd's pipe ?

Comatas. What manner of pipe might that
be, for when gat'st *thou* a pipe, thou slave of
Sibyrtas ? Why does it no more suffice thee
to keep a flute of straw, and whistle with
Corydon ?

Lacon. What pipe, free sir ? why, the pipe
that Lycon gave me. And what manner of
goat-skin hadst thou, that Lacon made off with?
Tell me, Comatas, for truly even thy master,
Eumarides, had never a goat-skin to sleep in.

Comatas. 'Twas the skin that Crocylus gave
me, the dappled one, when he sacrificed the
she-goat to the nymphs ; but thou, wretch,

even then wert wasting with envy, and now, at last, thou hast stripped me bare !

Lacon. Nay verily, so help me Pan of the seashore, it was not Lacon the son of Calaethis that filched the **coat of skin.** If I lie, **sirrah,** may I leap **frenzied down** this rock into **the** Crathis !

Comatas. Nay verily, my friend, so help me these nymphs of the mere (and ever may they be favourable, as now, and kind to me), it was not Comatas that pilfered thy pipe.

Lacon. If I believe thee, may I suffer the afflictions of Daphnis ! But see, if thou carest **to stake a** kid—though indeed 'tis **scarce** worth my while—then, go to, I will sing against thee, and cease not, till thou dost cry ' enough ! '

Comatas. The sow defied Athene ! See, there is staked the kid, go to, do thou too put a fatted lamb against him, for thy stake.

Lacon. Thou fox, and where would be our even betting then ? Who ever chose hair to shear, in place of wool ? and who prefers to milk a filthy bitch, when he can have a she-goat, nursing her first kid ?

Comatas. Why, he that **deems** himself as sure of getting the better of his neighbour as thou dost, a wasp that buzzes against the cicala. But as it is plain thou thinkst the kid no fair stake, lo, here is this he-goat. Begin the match !

Lacon. No such haste, thou art not on fire ! More sweetly wilt thou sing, if thou wilt sit down beneath the wild olive tree, and the groves in this place. Chill water falls there, drop by drop, here grows the grass, and here a leafy bed is strown, and here the locusts prattle.

Comatas. Nay, no whit am I in haste, but I am sorely vexed, that thou shouldst dare to look me straight in the face, thou whom I used to teach while thou wert still a child. See where gratitude goes! As well rear wolf-whelps, breed hounds, that they may devour thee!

Lacon. And what good thing have I to remember that I ever learned or heard from thee, thou envious thing, thou mere hideous manikin!

.　　.　　.　　.　　.

But come this way, come, and thou shalt sing thy last of country song.

Comatas. That way I will not go! Here be oak trees, and here the galingale, and sweetly here hum the bees about the hives. There are two wells of chill water, and on the tree the birds are warbling, and the shadow is beyond compare with that where thou liest, and from on high the pine tree pelts us with her cones.

Lacon. Nay, but lambs' wool, truly, and fleeces, shalt thou tread here, if thou wilt but come,—fleeces more soft than sleep, but the goat-skins beside thee stink—worse than thyself. And I will set a great bowl of white milk for the nymphs, and another will I offer of sweet olive oil.

Comatas. Nay, but an if thou wilt come,

thou shalt tread here the soft feathered fern, and flowering thyme, and beneath. thee shall be strown the skins of she-goats, four times more soft than **the fleeces of** thy lambs. And I will set out eight bowls **of** milk for Pan, and eight bowls full of the richest honeycombs.

Lacon. Thence, where thou art, I pray **thee,** begin the match, and there sing thy country song, tread thine own ground and keep thine oaks to thyself. But who, who shall judge between **us**? Would that Lycopas, the neat-herd, might chance to come this way !

Comatas. I want nothing with him, but that man, if thou wilt, that woodcutter we will call, who is **gathering those tufts** of heather near thee. It is Morson.

Lacon. Let us shout, then !

Comatas. Call thou to him.

Lacon. Ho, friend, come hither and listen for a little while, for we two have a match to prove which is the better singer of country song. So Morson, my friend, neither judge me too kindly, **no, nor** show him favour.

Comatas. Yes, dear Morson, **for the nymphs'** sake neither lean in thy judgment to Comatas, nor, prithee, favour *him.* The flock of sheep thou seest here belongs to Sibyrtas of Thurii, and the goats, friend, that thou beholdest are the goats of Eumarides of Sybaris.

Lacon. Now, in the name of Zeus did any one ask thee, thou make-mischief, who owned the flock, **I or** Sibyrtas? What a chatterer thou art !

Comatas. Best of men, I am for speaking the whole truth, and boasting never, but thou art too fond of cutting speeches.

Lacon. Come, say whatever thou hast to say, and let the stranger get home to the city alive; oh, Paean, what a babbler thou art, Comatas!

THE SINGING MATCH.

Comatas. The Muses love me better far than the minstrel Daphnis; but a little while ago I sacrificed two young she-goats to the Muses.

Lacon. Yea, and me too Apollo loves very dearly, and a noble ram I rear for Apollo, for the feast of the Carnea, look you, is drawing nigh.

Comatas. The she-goats that I milk have all borne twins save two. The maiden saw me, and 'alas,' she cried, 'dost thou milk alone?'

Lacon. Ah, ah, but Lacon here hath nigh twenty baskets full of cheese, and Lacon lies with his darling in the flowers!

Comatas. Clearista, too, pelts the goatherd with apples as he drives past his she-goats, and a sweet word she murmurs.

Lacon. And wild with love am I too, for my fair young darling, that meets the shepherd, with the bright hair floating round the shapely neck.

Comatas. Nay, ye may not liken dog-roses to the rose, or wind-flowers to the roses of the garden; by the garden walls their beds are blossoming.

Lacon. Nay, nor wild apples to acorns, for acorns are bitter in the oaken rind, but apples are sweet as honey.

Comatas. Soon will I give my maiden a ring-dove for a gift; I will take it from the juniper tree, for there it is brooding.

Lacon. But I will give my darling a soft fleece to make a cloak, a free gift, when I shear the black ewe.

Comatas. Forth from the wild olive, my bleating she-goats, feed here where the hillside slopes, and the tamarisks grow.

Lacon. Conarus there, and Cynaetha, will you never leave the oak? Graze here, where Phalarus feeds, where the hillside fronts the dawn.

Comatas. Ay, and I have a vessel of cypress wood, and a mixing bowl, the work of Praxiteles, and I hoard them for my maiden.

Lacon. I too have a dog that loves the flock, the dog to strangle wolves; him I am giving to my darling to chase all manner of wild beasts.

Comatas. Ye locusts that overleap our fence, see that ye harm not our vines, for our vines are young.

Lacon. Ye cicalas, see how I make the goatherd chafe : even so, methinks, do ye vex the reapers.

Comatas. I hate the foxes, with their bushy brushes, that ever come at evening, and eat the grapes of Micon.

Lacon. And I hate the lady-birds that

devour the figs of Philondas, and fiit down the wind.

Comatas. Dost thou not remember how I cudgelled thee, and thou didst grin and nimbly writhe, and catch hold of yonder oak ?

Lacon. **That** I have no memory of, but how Eumarides bound thee **there, upon a time,** and flogged **thee** through **and through, that** I do very well remember.

Comatas. Already, Morson, **some one** is waxing bitter, **dost** thou see no sign of it ? Go, go, and pluck, forthwith, the squills from some old wife's grave.

Lacon. And I too, Morson, I make some one chafe, and thou dost perceive it. Be off now to the Hales stream, and dig cyclamen.

Comatas. Let Himera flow with milk instead of water, and thou, **Crathis, run red with** wine, **and all** thy reeds **bear** apples.

Lacon. Would that the fount of Sybaris may flow with honey, and may the maiden's pail, at dawning, be dipped, not **in water, but** in **the** honeycomb.

Comatas. My goats eat cytisus, and goats-wort, **and** tread the lentisk shoots, and lie at ease among the arbutus.

Lacon. **But my ewes** have honey-wort to feed **on, and** luxuriant creepers flower around, as fair as roses.

Comatas. I love not Alcippe, for yesterday she did not kiss me, and take my face between her hands, when I gave her the dove.

Lacon. But deeply I love my darling, **for a**

kind kiss once I got, in return for the gift of a shepherd's pipe.

Comatas. Lacon, it never was right that pyes should contend with the nightingale, nor hoopoes with **swans**, but thou, unhappy swain, art ever for contention.

Morson's Judgment. I bid the shepherd cease. But to thee, Comatas, Morson presents **the** lamb. And thou, when thou hast sacrificed her to **the** nymphs, send Morson, anon, a goodly portion of her flesh.

Comatas. I will, by Pan. Now leap, and snort, my he-goats, all **the** herd of you, and see here how loud **I** ever will laugh, and exult over **Lacon, the** shepherd, for that, at last, I have won the lamb. See, I will leap sky high **with** joy. Take heart, my horned goats, to-morrow I will dip you **all** in the fountain of Sybaris. Thou white he-goat, I will beat thee if thou dare to touch one of the herd before I sacrifice the lamb to the nymphs. There he is at it again! Call me Melanthius,[1] not Comatas, if **I do not cudgel thee.**

[1] Melanthius was the treacherous goatherd put to a cruel death by Odysseus.

IDYL VI

Daphnis and Damoetas, two herdsmen of the golden age, meet by a well-side, and sing a match, their topic is the Cyclops, Polyphemus, and his love for the sea-nymph, Galatea.

The scene is in Sicily.

DAMOETAS, and Daphnis the herdsman, once on a time, Aratus, led the flock together into one place. Golden was the down on the chin of one, the beard of the other was half-grown, and by a well-head the twain sat them down, in the summer noon, and thus they sang. 'Twas Daphnis that began the singing, for the challenge had come from Daphnis.

Daphnis's Song of the Cyclops.

Galatea is pelting thy flock with apples, Polyphemus, she says the goatherd is a laggard lover! And thou dost not glance at her, oh hard, hard that thou art, but still thou sittest at thy sweet piping. Ah see, again, she is pelting thy dog, that follows thee to watch thy sheep. He barks, as he looks into the brine, and now the beautiful waves that softly plash

reveal him,[1] as he runs upon the shore. Take
heed that he leap not on the maiden's limbs as
she rises from the salt water, see that he rend
not her lovely body! Ah, thence again, see,
she is wantoning, light as dry thistle-down in
the scorching summer weather. She flies when
thou art wooing her ; when thou woo'st not she
pursues thee, she plays out all her game and
leaves her king unguarded. For truly to Love,
Polyphemus, many a time doth foul seem fair!

*He ended, and Damoetas touched a prelude
to his sweet song.*

I saw her, by Pan, I saw her when she was
pelting my flock. Nay, she escaped not me,
escaped not my one dear eye,—wherewith I
shall see to my life's end,—let Telemus the
soothsayer, that prophesies hateful things, hateful things take home, to keep them for his
children! But it is all to torment her, that
I, in my turn, give not back her glances, pre-
tending that I have another love. To hear
this makes her jealous of me, by Paean, and
she wastes with pain, and springs madly from
the sea, gazing at my caves and at my herds.
And I hiss on my dog to bark at her, for when
I loved Galatea he would whine with joy, and
lay his muzzle on her lap. Perchance when
she marks how I use her she will send me
many a messenger, but on her envoys I will

---

[1] Ameis and Fritzsche take νιν (as here) to be the
dog, not Galatea. The sex of the Cyclops's sheep-dog
makes the meaning obscure.

shut my door till she promises that herself will make a glorious bridal-bed **on this** island for me. For in truth, I am not so hideous as they say! But lately I was looking into the **sea,** when all was calm ; beautiful seemed my **beard,** beautiful my one eye—as I count beauty—and the sea reflected the gleam **of** my teeth whiter than the Parian stone. Then, all **to shun the evil eye, did I spit thrice in** my **breast** ; for this **spell** was taught me by the crone, Cottytaris, that piped of yore to the reapers in Hippocoon's field.

Then Damoetas kissed Daphnis, as he ended his song, and he gave Daphnis a pipe, and Daphnis gave him a beautiful flute. Damoetas fluted, and Daphnis piped, the herdsman,—and anon the calves were dancing in the soft green grass. Neither won the victory, but both were invincible.

IDYL VII

*The poet making his way through the noonday heat,
with two friends, to a harvest feast, meets the goatherd, Lycidas. To humour the poet, Lycidas sings
a love song of his own, and the other replies with
verses about the passion of Aratus, the famous writer
of didactic verse. After a courteous parting from
Lycidas, the poet and his two friends repair to the
orchard, where Demeter is being gratified with the
first-fruits of harvest and vintaging.*

*In this idyl, Theocritus, speaking of himself by the name
of Simichidas, alludes to his teachers in poetry,
and, perhaps, to some of the literary quarrels of the
time.*

*The scene is in the isle of Cos. G. Hermann fancied
that the scene was in Lucania, and Mr. W. R.
Paton thinks he can identify the places named by
the aid of inscriptions* (Classical Review, ii. 8,
265). *See also Rayet,* Mémoire sur l'île de Cos,
p. 18, *Paris,* 1876.

The Harvest Feast.

IT fell upon a time when Eucritus and I were
walking from the city to the Hales water, and
Amyntas was the third in our company. The
harvest-feast of Deo was then being held by
Phrasidemus and Antigenes, two sons of Lyco-
peus (if aught there be of noble and old descent),

whose lineage dates from Clytia, and Chalcon
himself—Chalcon, beneath whose foot the fountain sprang, the well of Buriné. He set his
knee stoutly against the rock, and straightway
by the spring poplars and elm trees showed
a shadowy glade, arched overhead they grew,
and pleached with leaves of green. We had
not yet reached the mid-point of the way, nor
was the tomb of Brasilas yet risen upon our
sight, when,—thanks be to the Muses—we met
a certain wayfarer, the best of men, a Cydonian.
Lycidas was his name, a goatherd was he, nor
could any that saw him have taken him for
other than he was, for all about him bespoke
the goatherd. Stripped from the roughest of
he-goats was the tawny skin he wore on his
shoulders, the smell of rennet clinging to it
still, and about his breast an old cloak was
buckled with a plaited belt, and in his right
hand he carried a crooked staff of wild olive ;
and quietly he accosted me, with a smile, a
twinkling eye, and a laugh still on his lips :—

'Simichidas, whither, pray, through the noon
dost thou trail thy feet, when even the very
lizard on the rough stone wall is sleeping, and
the crested larks no longer fare afield? Art
thou hastening to a feast, a bidden guest, or
art thou for treading a townsman's wine-press?
For such is thy speed that every stone upon
the way spins singing from thy boots !'

'Dear Lycidas,' I answered him, 'they all
say that thou among herdsmen, yea, and reapers
art far the chiefest flute-player. In sooth this

greatly rejoices our hearts, and yet, to my conceit, meseems I can vie with thee. But as to this journey, we are going to the harvest-feast, for, look you some friends of ours are paying a festival to fair-robed Demeter, out of the first-fruits of their increase, for verily in rich measure has the goddess filled their threshing-floor with barley grain. But come, for the way and the day are thine alike and mine, come, let us vie in pastoral song, perchance each will make the other delight. For I, too, am a clear-voiced mouth of the Muses, and they all call me the best of minstrels, but I am not so credulous ; no, by Earth, for to my mind I cannot as yet conquer in song that great Sicelidas — the Samian—nay, nor yet Philetas. 'Tis a match of frog against cicala !'

So I spoke, to win my end, and the goatherd with his sweet laugh, said, ' I give thee this staff, because thou art a sapling of Zeus, and in thee is no guile. For as I hate your builders that try to raise a house as high as the mountain summit of Oromedon,[1] so I hate all birds of the Muses that vainly toil with their cackling notes against the Minstrel of Chios ! But come, Simichidas, without more ado let us begin the pastoral song. And I—nay, see friend—if it please thee at all, this ditty that I lately fashioned on the mountain side !'

---

[1] Or, ὅμον Ὠρομέδοντος. Hermann renders this *domum Oromedonteam* 'a gigantic house.' Oromedon or Eurymedon was the king of the Gigantes, mentioned in Odyssey vii. 58.

The Song of Lycidas.

Fair voyaging befall Ageanax to Mytilene, both when the *Kids* are westering, and the south wind the wet waves chases, and when Orion holds his feet above the Ocean! Fair voyaging betide him, if he saves Lycidas from the fire of Aphrodite, for hot is the love that consumes me.

The halcyons will lull the waves, and lull the deep, and the south wind, and the east, that stirs the sea-weeds on the farthest shores,[1] the halcyons that are dearest to the green-haired mermaids, of all the birds that take their prey from the salt sea. Let all things smile on Ageanax to Mytilene sailing, and may he come to a friendly haven. And I, on that day, will go crowned with anise, or with a rosy wreath, or a garland of white violets, and the fine wine of Ptelea I will dip from the bowl as I lie by the fire, while one shall roast beans for me, in the embers. And elbow-deep shall the flowery bed be thickly strewn, with fragrant leaves and with asphodel, and with curled parsley; and softly will I drink, toasting Ageanax with lips clinging fast to the cup, and draining it even to the lees.

Two shepherds shall be my flute-players, one from Acharnae, one from Lycope, and hard by

---

[1] ἔσχατα. This is taken by some to mean *algam infimam*, 'the bottom weeds of the deepest seas,' by others, the sea-weed highest on the shore, at high water-mark.

Tityrus shall sing, how the herdsman Daphnis
once loved a strange maiden, and how on
the hill he wandered, and how the oak trees
sang his dirge—the oaks that grow by the
banks of the river Himeras—while he was
wasting like any snow under high Haemus,
or Athos, or Rhodope, or Caucasus at the
world's end.

And he shall sing how, once upon a time,
the great chest prisoned the living goatherd, by
his lord's infatuate and evil will, and how the
blunt-faced bees, as they came up from the
meadow to the fragrant cedar chest, fed him
with food of tender flowers, because the Muse
still dropped sweet nectar on his lips.[1]

O blessed Comatas, surely these joyful
things befell thee, and thou wast enclosed
within the chest, and feeding on the honeycomb through the springtime didst thou serve
out thy bondage. Ah, would that in my days
thou hadst been numbered with the living, how
gladly on the hills would I have herded thy
pretty she-goats, and listened to thy voice,
whilst thou, under oaks or pine trees lying,
didst sweetly sing, divine Comatas!

When he had chanted thus much he ceased,

---

[1] Comatas was a goatherd who devoutly served the
Muses, and sacrificed to them his master's goats. His
master therefore shut him up in a cedar chest, opening
which at the year's end he found Comatas alive, by
miracle, the bees having fed him with honey. Thus, in
a mediaeval legend, the Blessed Virgin took the place,
for a year, of the frail nun who had devoutly served
her

and I followed after him again, with some such words as these :—

'Dear Lycidas, many another song the Nymphs have taught me also, as I followed my herds upon the hillside, bright songs that Rumour, perchance, has brought even to the throne of Zeus. But of them all this is far the most excellent, **wherewith** I will begin to do thee honour : nay listen as thou art dear to the **Muses.**'

The Song of Simichidas.

For Simichidas the Loves have sneezed, for truly the wretch loves Myrto as dearly as goats love the spring.[1] But Aratus, far the dearest of my friends, deep, deep in his heart he keeps Desire,—and Aratus's love is young ! Aristis knows it, an honourable man, nay of men the best, whom even Phoebus would permit to stand and sing lyre in hand, by his tripods. Aristis knows how deeply love is burning Aratus to the bone. Ah, Pan, thou lord of the beautiful plain of Homole, bring, I pray thee, the darling of Aratus unbidden to his arms, whosoe'er it be that he loves. If this thou dost, dear Pan, then never may the boys of Arcady flog thy sides and shoulders with stinging herbs, when scanty meats are left them on thine altar. But if thou shouldst otherwise decree, then may all thy skin be frayed and torn with thy nails, yea, and in nettles mayst

---

[1] Sneezing in Sicily, as in most countries, was a happy omen.

thou couch! In the hills of the Edonians
mayst thou dwell in mid-winter time, by the
river Hebrus, close neighbour to the Polar
star! But in summer mayst thou range with
the uttermost Æthiopians beneath the rock of
the Blemyes, whence Nile no more is seen.

And you, leave ye the sweet fountain of
Hyetis and Byblis, and ye that dwell in the
steep home of golden Dione, ye Loves as rosy
as red apples, strike me with your arrows, the
desired, the beloved ; strike, for that ill-starred
one pities not my friend, my host! And yet
assuredly the pear is over-ripe, and the maidens
cry ' alas, alas, thy fair bloom fades away!'

Come, no more let us mount guard by these
gates, Aratus, nor wear our feet away with
knocking there. Nay, let the crowing of the
morning cock give others over to the bitter
cold of dawn. Let Molon alone, my friend,
bear the torment at that school of passion!
For us, let us secure a quiet life, and some old
crone to spit on us for luck, and so keep all
unlovely things away.

Thus I sang, and sweetly smiling, as before,
he gave me the staff, a pledge of brotherhood
in the Muses. Then he bent his way to the
left, and took the road to Pyxa, while I and
Eucritus, with beautiful Amyntas, turned to the
farm of Phrasidemus. There we reclined on
deep beds of fragrant lentisk, lowly strown,
and rejoicing we lay in new stript leaves of 'the
vine. And high above our heads waved many
a poplar, many an elm tree, while close at hand

the sacred water from the nymphs' own cave
welled forth with murmurs musical. On
shadowy boughs the burnt cicalas kept their
chattering toil, far off the little owl cried in the
thick thorn brake, the larks and finches were
singing, the ring-dove moaned, the yellow bees
were flitting about the springs. All breathed
the scent of the opulent summer, of the season
of fruits ; pears at our feet and apples by our
sides were rolling plentiful, the tender branches,
with wild plums laden, were earthward bowed,
and the four-year-old pitch seal was loosened
from the mouth of the wine-jars.

Ye nymphs of Castaly that hold the steep of
Parnassus, say, was it ever a bowl like this that
old Chiron set before Heracles in the rocky
cave of Pholus ? Was it nectar like this that
beguiled the shepherd to dance and foot it
about his folds, the shepherd that dwelt by
Anapus, on a time, the strong Polyphemus
who hurled at ships with mountains ? Had
these ever such a draught as ye nymphs bade
flow for us by the altar of Demeter of the
threshing-floor ?

Ah, once again may I plant the great fan on
her corn-heap, while she stands smiling by, with
sheaves and poppies in her hands.

IDYL VIII

The scene is among the high mountain pastures of Sicily :—

> ' *On the sward, at the cliff top*
> *Lie strewn the white flocks ;* '

and far below shines and murmurs the Sicilian sea. Here Daphnis and Menalcas, two herdsmen of the golden age, meet, while still in their earliest youth, and contend for the prize of pastoral. Their songs, in elegiac measure, are variations on the themes of love and friendship (for Menalcas sings of Milon, Daphnis of Nais), and of nature. Daphnis is the winner; it is his earliest victory, and the prelude to his great renown among nymphs and shepherds. In this version the strophes are arranged as in Fritzsche's text. Some critics take the poem to be a patchwork by various hands.

As beautiful Daphnis was following his kine, and Menalcas shepherding his flock, they met, as men tell, on the long ranges of the hills. The beards of both had still the first golden bloom, both were in their earliest youth, both were pipe-players skilled, both skilled in song. Then first Menalcas, looking at Daphnis, thus bespoke him.

' Daphnis, thou herdsman of the lowing kine,

art thou minded to sing a match with me?
Methinks I shall vanquish thee, when I sing in
turn, as readily as I please.'

Then Daphnis answered him again in this
wise, 'Thou shepherd of the fleecy sheep,
Menalcas, the pipe-player, never wilt thou
vanquish **me in song,** not thou, if thou **shouldst**
sing till some evil thing befall thee !'

Menalcas. **Dost thou** care then, to **try this**
and see, dost thou care to risk a stake ?

Daphnis. I do care to try this and see, a
stake I am ready to risk.

Menalcas. But what shall we stake, what
pledge shall we find equal and sufficient ?

Daphnis. I will pledge a calf, and do thou
put down a lamb, one that has grown to his
mother's height.

Menalcas. **Nay,** never will **I stake a lamb,**
for stern is my father, and stern **my mother,**
and they number all the sheep at **evening.**

Daphnis. But what, then, wilt thou lay, and
where is to be the victor's gain ?

Menalcas. The pipe, the fair pipe with nine
stops, that I made myself, fitted with white
wax, and smoothed evenly, above as below.
This would I readily wager, **but never** will I
stake aught that is my father's.

Daphnis. See then, I too, in truth, have a
pipe with nine stops, fitted with white wax,
and smoothed evenly, above as below. But
lately I put it together, and this finger
still aches, where the reed split, and cut it
deeply.

Menalcas. But who is to judge between us, who will listen to our singing ?

Daphnis. That goatherd yonder, he will do, if we call him hither, the man for whom that dog, a black hound with a white patch, is barking among the kids.

Then the boys called aloud, and the goatherd gave ear, and came, and the boys began to sing, and the goatherd was willing to be their umpire. And first Menalcas sang (for he drew the lot) the sweet-voiced Menalcas, and Daphnis took up the answering strain of pastoral song—and 'twas thus Menalcas began :

Menalcas. Ye glades, ye rivers, issue of the Gods, if ever Menalcas the flute-player sang a song ye loved, to please him, feed his lambs ; and if ever Daphnis come hither with his calves, may he have no less a boon.

Daphnis. Ye wells and pastures, sweet growth o' the world, if Daphnis sings like the nightingales, do ye fatten this herd of his, and if Menalcas hither lead a flock, may he too have pasture ungrudging to his full desire !

Menalcas. There doth the ewe bear twins, and there the goats ; there the bees fill the hives, and there oaks grow loftier than common, wheresoever beautiful Milon's feet walk wandering ; ah, if he depart, then withered and lean is the shepherd, and lean the pastures !

Daphnis. Everywhere is spring, and pastures everywhere, and everywhere the cows' udders are swollen with milk, and the younglings are fostered, wheresoever fair Nais roams ; ah, if she depart, then parched are the kine, and he that feeds them !

Menalcas. O bearded goat, thou mate of the white herd, and O ye blunt-faced kids, where are the manifold deeps of the forest, thither get ye to the water, for thereby is Milon ; go, thou hornless goat, and say to him, ‘ Milon, Proteus was a herdsman, and that of seals, though he was a god.’

Daphnis.   .   .   .   .   .

Menalcas. Not mine be the land of Pelops, not mine to own talents of gold, nay, nor mine to outrun the speed of the winds ! Nay, but beneath this rock will I sing, with thee in mine arms, and watch our flocks feeding together, and, before us, the Sicilian sea.

Daphnis.   .   .   .   .   .

Menalcas.   .   .   .   .   .

Daphnis. Tempest is the dread pest of the trees, drought of the waters, snares of the birds, and the hunter's net of the wild beasts, but ruinous to man is the love of a delicate maiden. O father, O Zeus, I have not been the only lover, thou too hast longed for a mortal woman.

Thus the boys sang in verses amoebaean, and thus Menalcas began the crowning lay :

Menalcas. Wolf, spare the kids, spare the mothers of my herd, and harm not me, so young as I am to tend so great a flock. Ah, Lampurus, my dog, dost thou then sleep so soundly ? a dog should not sleep so sound, that helps a boyish shepherd. Ewes of mine, spare ye not to take your fill of the tender herb, ye shall not weary, 'ere all this grass grows again. Hist, feed on, feed on, fill, all of you, your udders, that there may be milk for the lambs, and somewhat for me to store away in the cheese-crates.

Then Daphnis followed again, and sweetly preluded to his singing :

Daphnis. Me, even me, from the cave, the girl with meeting eyebrows spied yesterday as I was driving past my calves, and she cried, ' How fair, how fair he is ! ' But I answered her never the word of railing, but cast down my eyes, and plodded on my way.

Sweet is the voice of the heifer, sweet her breath,[1] sweet to lie beneath the sky in summer, by running water.

Acorns are the pride of the oak, apples of the apple tree, the calf of the heifer, and the neatherd glories in his kine.

So sang the lads, and the goatherd thus bespoke them, ' Sweet is thy mouth, O Daphnis, and delectable thy song ! Better is it to listen to thy singing, than to taste the honeycomb. Take thou the pipe, for thou hast conquered in the singing match. Ah, if thou wilt but teach some lay, even to me, as I tend the goats beside thee, this blunt-horned she-goat will I give thee, for the price of thy teaching, this she-goat that ever fills the milking pail above the brim. '

Then was the boy as glad,—and leaped high, and clapped his hands over his victory, —as a young fawn leaps about his mother.

---

[1] A superfluous and apocryphal line is here omitted.

But the heart of the other was wasted with grief, and desolate, even as a maiden sorrows that is newly wed.

From this time Daphnis became the foremost among the shepherds, and while yet in his earliest youth, he wedded the nymph Nais.

IDYL IX

Daphnis and Menalcas, at the bidding of the poet, sing the joys of the neatherd's and of the shepherd's life. Both receive the thanks of the poet, and rustic prizes — a staff, and a horn, made of a spiral shell. Doubts have been expressed as to the authenticity of the prelude and concluding verses. The latter breathe all Theocritus's enthusiastic love of song.

SING, Daphnis, a pastoral lay, do thou first begin the song, the song begin, O Daphnis; but let Menalcas join in the strain, when ye have mated the heifers and their calves, the barren kine and the bulls. Let them all pasture together, let them wander in the coppice, but never leave the herd. Chant thou for me, first, and on the other side let Menalcas reply.

Daphnis. Ah, sweetly lows the calf, and sweetly the heifer, sweetly sounds the neatherd with his pipe, and sweetly also I! My bed of leaves is strown by the cool water, and thereon are heaped fair skins from the white calves that were all browsing upon the arbutus, on a time, when the south-west wind dashed me them from the height.

And thus I heed no more the scorching summer, than a lover cares to heed the words of father or of mother.

So Daphnis sang to me, and thus, in turn, did Menalcas sing.

Menalcas. Aetna, mother mine, I too dwell in a beautiful cavern in the chamber of the rock, and, lo, all the wealth have I that we behold in dreams; ewes in plenty and she-goats abundant, their fleeces are strown beneath my head and feet. In the fire of oak-faggots puddings are hissing-hot, and dry beech-nuts roast therein, in the wintry weather, and, truly, for the winter season I care not even so much as a toothless man does for walnuts, when rich pottage is beside him.

Then I clapped my hands in their honour, and instantly gave each a gift, to Daphnis a staff that grew in my father's close, self-shapen, yet so straight, that perchance even a craftsman could have found no fault in it. To the other I gave a goodly spiral shell, the meat that filled it once I had eaten after stalking the fish on the Icarian rocks (I cut it into five shares for five of us),—and Menalcas blew a blast on the shell.

Ye pastoral Muses, farewell! Bring ye into the light the song that I sang there to these shepherds on that day! Never let the pimple grow on my tongue-tip.[1]

---

[1] An allusion to the common superstition (cf. Idyl xii. 24) that perjurers and liars were punished by pimples and blotches. The old Irish held that blotches showed

Cicala to cicala is dear, and ant to ant, and hawks to hawks, but to me the Muse and song. Of song may all my dwelling be full, for sleep is not more sweet, nor sudden spring, nor flowers are more delicious to the bees—so dear to me are the Muses.[1] Whom they look on in happy hour, Circe hath never harmed with her enchanted potion.

themselves on the faces of Brehons who gave unjust judgments.

[1] Spring in the south, like Night in the tropics, comes 'at one stride'; but Wordsworth finds the rendering distasteful, 'neque sic redditum valde placet.'

THE REAPERS

This is an idyl of the same genre *as Idyl IV. The sturdy reaper, Milon, as he levels the swathes of corn, derides his languid and love-worn companion, Battus. The latter defends his gipsy love in verses which have been the keynote of much later poetry, and which echo in the fourth book of Lucretius, and in the* Misanthrope *of Molière. Milon replies with the song of Lityerses—a string, apparently, of popular rural couplets, such as Theocritus may have heard chanted in the fields.*

Milon. Thou **toilsome** clod; **what** ails thee now, thou wretched fellow? Canst thou neither cut thy swathe straight, as thou wert wont to do, nor keep time with thy neighbour **in thy** reaping, **but** thou **must** fall out, like an **ewe** that is foot-pricked with a thorn and straggles **from** the herd? What manner of man wilt thou prove after mid-noon, and at evening, thou that dost not prosper with thy swathe when **thou art** fresh begun?

Battus. Milon, thou that canst toil till late, thou chip of the stubborn stone, has it never befallen thee to long for one that was not with thee?

Milon. Never! What has a labouring man
to do with hankering after what he has not
got?

Battus. Then it never befell thee to lie
awake for love?

Milon. Forbid it ; 'tis an ill thing to let the
dog once taste of pudding.

Battus. But I, Milon, am in love for almost
eleven days !

Milon. 'Tis easily seen that thou drawest
from a wine-cask, while even vinegar is scarce
with me.

Battus. And for Love's sake, the fields before
my doors are untilled since seed-time.

Milon. But which of the girls afflicts thee so?

Battus. The daughter of Polybotas, she that
of late was wont to pipe to the reapers on
Hippocoon's farm.

Milon. God has found out the guilty ! Thou
hast what thou'st long been seeking, that grasshopper of a girl will lie by thee the night long !

Battus. Thou art beginning thy mocks of
me, but Plutus is not the only blind god ; he
too is blind, the heedless Love ! Beware of
talking big.

Milon. Talk big I do not ! Only see that
thou dost level the corn, and strike up some
love-ditty in the wench's praise. More plea-
santly thus wilt thou labour, and, indeed, of
old thou wert a melodist.

Battus. Ye Muses Pierian, sing ye with me
the slender maiden, for whatsoever ye do but
touch, ye goddesses, ye make wholly fair.

They all call thee a *gipsy*, gracious Bombyca, and *lean*, and *sunburnt*, 'tis only I that call thee *honey-pale*.

Yea, and the violet is swart, and swart the lettered hyacinth, but yet these flowers are chosen the first in garlands.

The goat runs after cytisus, the wolf pursues the goat, the crane follows the plough, but I am wild for love of thee.

Would it were mine, all the wealth whereof once Croesus was lord, as men tell! Then images of us twain, all in gold, should be dedicated to Aphrodite, thou with thy flute, and a rose, yea, or an apple, and I in fair attire, and new shoon of Amyclae on both my feet.

Ah gracious Bombyca, thy feet are fashioned like carven ivory, thy voice is drowsy sweet, and thy ways, I cannot tell of them![1]

Milon. Verily our clown was a maker of lovely songs, and we knew it not! How well he meted out and shaped his harmony; woe is me for the beard that I have grown, all in vain! Come, mark thou too these lines of godlike Lityerses!

THE LITYERSES SONG.

Demeter, rich in fruit, and rich in grain, may this corn be easy to win, and fruitful exceedingly!

Bind, ye bandsters, the sheaves, lest the way-

[1] 'Quant à ta manière, je ne puis la rendre.'—
SAINTE-BEUVE.

*farer should cry, 'Men of straw were the
workers here, ay, and their hire was wasted!'*

*See that the cut stubble faces the North wind,
or the West, 'tis thus the grain waxes richest.*

*They that thresh corn should shun the noon-
day sleep; at noon the chaff parts easiest from
the straw.*

*As for the reapers, let them begin when the
crested lark is waking, and cease when he sleeps,
but take holiday in the heat.*

*Lads, the frog has a jolly life, he is not
cumbered about a butler to his drink, for he has
liquor by him unstinted!*

*Boil the lentils better, thou miserly steward;
take heed lest thou chop thy fingers, when thou'rt
splitting cumin-seed.*

'Tis thus that men should sing who labour
i' the sun, but thy starveling love, thou clod,
'twere fit to tell to thy mother when she stirs in
bed at dawning.

IDYL XI

THE CYCLOPS IN LOVE

*Nicias, the physician and poet, being in love, Theocritus
reminds him that in song lies the only remedy. It
was by song, he says, that the Cyclops, Polyphemus,
got him some ease, when he was in love with Galatea,
the sea-nymph.*

*The idyl displays, in the most graceful manner, the Alex-
andrian taste for turning Greek mythology into love
stories. No creature could be more remote from love
than the original Polyphemus, the cannibal giant
of the Odyssey.*

THERE is none other medicine, Nicias, against
Love, neither unguent, methinks, nor salve to
sprinkle,—none, save the Muses of Pieria!
Now a delicate thing is their minstrelsy in man's
life, and a sweet, but hard to procure. Methinks
thou know'st this well, who art thyself a leech,
and beyond all men art plainly dear to the
Muses nine.

'Twas surely thus the Cyclops fleeted his life
most easily, he that dwelt among us,—Polyphemus of old time,—when the beard was yet
young on his cheek and chin ; and he loved
Galatea. He loved, not with apples, not roses,

nor locks of hair, but with fatal frenzy, and all
things else he held but trifles by the way.
Many a time from the green pastures would his
ewes stray back, self-shepherded, to the fold.
But he was singing of Galatea, and pining in
his place he sat by the sea-weed of the beach,
from the dawn of day, with the direst hurt
beneath his breast of mighty Cypris's sending,
—the wound of her arrow in his heart !

Yet this remedy he found, and sitting on the
crest of the tall cliff, and looking to the deep,
'twas thus he would sing :—

Song of the Cyclops.

O milk-white Galatea, why cast off him
that loves thee ?   More white than is pressed
milk to look upon, more delicate than the
lamb art thou, than the young calf wantoner,
more sleek than the unripened grape !   Here
dost thou resort, even so, when sweet sleep
possesses me, and home straightway dost thou
depart when sweet sleep lets me go, fleeing
me like an ewe that has seen the grey wolf.

I fell in love with thee, maiden, I, on the
day when first thou camest, with my mother, and
didst wish to pluck the hyacinths from the hill,
and I was thy guide on the way.   But to leave
loving thee, when once I had seen thee, neither
afterward, nor now at all, have I the strength,
even from that hour.   But to thee all this is as
nothing, by Zeus, nay, nothing at all !

I know, thou gracious maiden, why it is

that thou dost shun me. It is all for the
shaggy brow that spans all my forehead, from
this to the other ear, one long unbroken eye-
brow. And but one eye is on my forehead,
and broad is the nose that overhangs my lip.
Yet I (even such as thou seest me) feed a
thousand cattle, and from these I draw and
drink the best milk in the world. And cheese
I never lack, in summer time or autumn, nay,
nor in the dead of winter, but my baskets are
always overladen.

Also I am skilled in piping, as none other
of the Cyclopes here, and of thee, my love, my
sweet-apple, and of myself too I sing, many
a time, deep in the night. And for thee I tend
eleven fawns, all crescent-browed,[1] and four
young whelps of the bear.

Nay, come thou to me, and thou shalt lack
nothing that now thou hast. Leave the grey
sea to roll against the land ; more sweetly, in
this cavern, shalt thou fleet the night with me !
Thereby the laurels grow, and there the slender
cypresses, there is the ivy dun, and the sweet
clustered grapes ; there is chill water, that for
me deep-wooded Ætna sends down from the
white snow, a draught divine ! Ah who, in
place of these, would choose the sea to dwell
in, or the waves of the sea ?

But if thou dost refuse because my body
seems shaggy and rough, well, I have faggots
of oakwood, and beneath the ashes is fire un-
wearied, and I would endure to let thee burn

---

[1] Reading μηνοφόρως.

my very soul, and this my one eye, the dearest
thing that is mine.

Ah me, that my mother bore me not a finny
thing, so would I have gone down to thee, and
kissed thy hand, if thy lips thou would not
suffer me to kiss ! And I would have brought
thee either white lilies, or the soft poppy with
its scarlet petals. Nay, these are summer's
flowers, and those are flowers of winter, so I
could not have brought thee them all at one
time.

Now, verily, maiden, now and here will I
learn to swim, if perchance some stranger come
hither, sailing with his ship, that I may see
why it is so dear to thee, to have thy dwelling
in the deep.

Come forth, Galatea, and forget as thou
comest, even as I that sit here have forgotten,
the homeward way ! Nay, choose with me to
go shepherding, with me to milk the flocks, and
to pour the sharp rennet in, and to fix the
cheeses.

There is none that wrongs me but that
mother of mine, and her do I blame. Never,
nay, never once has she spoken a kind word
for me to thee, and that though day by day she
beholds me wasting. I will tell her that my
head, and both my feet are throbbing, that
she may somewhat suffer, since I too am
suffering.

O Cyclops, Cyclops, whither are thy wits
wandering? Ah that thou wouldst go, and
weave thy wicker - work, and gather broken

boughs to carry to thy lambs : in faith, if thou didst this, far wiser wouldst thou be !

Milk the ewe that thou hast, why pursue the thing that shuns thee ? Thou wilt find, perchance, another, and a fairer Galatea. Many be the girls that bid me play with them through the night, and softly they all laugh, if perchance I answer them. On land it is plain that I too seem to be somebody !

Lo, thus Polyphemus still shepherded his love with song, and lived lighter than if he had given gold for ease.

IDYL XII

THE PASSIONATE FRIEND

*This is rather a lyric than an idyl, being an expression
of that singular passion which existed between men
in historical Greece. The next idyl, like the Myr-
midons of Aeschylus, attributes the same manners
to mythical and heroic Greece. It should be un-
necessary to say, that the affection between Homeric
warriors, like Achilles and Patroclus, was only
that of companions in arms and was quite unlike
the later sentiment.*

HAST thou come, dear youth, with the third
night and the dawning; hast thou come? but
men in longing grow old in a day! As spring
than the winter is sweeter, as the apple than
the sloe, as the ewe is deeper of fleece than the
lamb she bore; as a maiden surpasses a thrice-
wedded wife, as the fawn is nimbler than the
calf; nay, by as much as sweetest of all fowls
sings the clear-voiced nightingale, so much
has thy coming gladdened me! To thee
have I hastened as the traveller hastens
under the burning sun to the shadow of the
ilex tree.

Ah, would that equally the Loves may breathe upon us twain, may we become a song in the ears of all men unborn.

'Lo, a pair were these two friends among the folk of former time,' the one 'the Knight' (so the Amyclaeans call him), the other, again, 'the Page,' so styled in speech of Thessaly.

'An equal yoke of friendship they bore : ah, surely then there were golden men of old, when friends gave love for love !'

And would, O father Cronides, and would, ye ageless immortals, that this might be ; and that when two hundred generations have sped, one might bring these tidings to me by Acheron, the irremeable stream.

'The loving-kindness that was between thee and thy gracious friend, is even now in all men's mouths, and chiefly on the lips of the young.'

Nay, verily, the gods of heaven will be masters of these things, to rule them as they will, but when I praise thy graciousness no blotch that punishes the perjurer shall spring upon the tip of my nose ! Nay, if ever thou hast somewhat pained me, forthwith thou healest the hurt, giving a double delight, and I depart with my cup full and running over !

Nisaean men of Megara, ye champions of the oars, happily may ye dwell, for that ye honoured above all men the Athenian stranger, even Diocles, the true lover. Always about his tomb the children gather in their companies, at the coming in of the spring, and contend for the prize of kissing. And whoso most sweetly touches lip to lip, laden with garlands he returneth to his mother. Happy is he that judges those kisses of the children ; surely he prays most earnestly to bright-faced Ganymedes, that his lips may be as the Lydian touchstone, wherewith the money - changers try gold lest perchance base metal pass for true.

HYLAS AND HERACLES

*As in the eleventh Idyl, Nicias is again addressed, by
way of introduction to the story of Hylas. This
beautiful lad, a favourite companion of Heracles,
took part in the Quest of the Fleece of Gold. As he
went to draw water from a fountain, the water-
nymphs dragged him down to their home, and
Heracles, after a long and vain search, was compelled to follow the heroes of the Quest on foot to
Phasis.*

NOT for us only, Nicias, as we were used to
deem, was Love begotten, by whomsoever of
the Gods was the father of the child ; not first
to us seemed beauty beautiful, to us that are
mortal men and look not on the morrow. Nay,
but the son of Amphitryon, that heart of bronze,
who abode the wild lion's onset, loved a lad,
beautiful Hylas—Hylas of the braided locks,
and he taught him all things as a father teaches
his child, all whereby himself became a mighty
man, and renowned in minstrelsy. Never was
he apart from Hylas, not when midnoon was
high in heaven, not when Dawn with her white

horses speeds upwards to the dwelling of Zeus, not when the twittering nestlings look towards the perch, while their mother flaps her wings above the smoke-browned beam ; and all this that the lad might be fashioned to his mind, and might drive a straight furrow, and come to the true measure of man.

But when Iason, Aeson's son, was sailing after the fleece of gold (and with him followed the champions, the first chosen out of all the cities, they that were of most avail), to rich Iolcos too came the mighty man and adventurous, the son of the woman of Midea, noble Alcmene.   With him went down Hylas also, to Argo of the goodly benches, the ship that grazed not on the clashing rocks Cyanean, but through she sped and ran into deep Phasis, as an eagle over the mighty gulf of the sea.   And the clashing rocks stand fixed, even from that hour !

Now at the rising of the Pleiades, when the upland fields begin to pasture the young lambs, and when spring is already on the wane, then the flower divine of Heroes bethought them of sea-faring.   On board the hollow Argo they sat down to the oars, and to the Hellespont they came when the south wind had been for three days blowing, and made their haven within Propontis, where the oxen of the Cianes wear bright the ploughshare, as they widen the furrows.   Then they went forth upon the shore, and each couple busily got ready supper in the late evening, and many as they were one bed

they strewed lowly on the ground, for they found
a meadow lying, rich in couches of strown grass
and leaves. Thence they cut them pointed
flag-leaves, and deep marsh-galingale. And
Hylas of the yellow hair, with a vessel of bronze
in his hand, went to draw water against supper-
time, for Heracles himself, and **the** steadfast
Telamon, for these comrades twain **supped ever**
at one table. Soon **was he ware** of a spring,
in a hollow land, and the rushes grew thickly
round it, and dark swallow-wort, and green
maiden-hair, and blooming parsley, and deer-
grass spreading through the marshy land. In
the midst of the water the nymphs were arraying
their dances, the sleepless nymphs, dread goddesses of the country people, Eunice, and Malis,
and Nycheia, with her April eyes. And now
the boy was holding out the wide-mouthed
pitcher to the water, intent on dipping **it, but**
the nymphs all clung to his hand, for love of
the Argive lad had **fluttered** the soft hearts of
all of them. Then down he sank into the
black water, headlong all, as when a star shoots
flaming from the sky, plumb in the deep it
falls, and a **mate** shouts out to the seamen,
' Up with the gear, my lads, the wind is fair for
sailing.'

Then the nymphs held the weeping boy **on**
their laps, and with gentle words were striving
to comfort him. But the son of Amphitryon
was troubled about the lad, and went forth,
carrying his bended bow in Scythian fashion,
and the club that is **ever grasped in his right**

hand. Thrice he shouted 'Hylas!' as loud as his deep throat could call, and thrice again the boy heard him, and thin came his voice from the water, and, hard by though he was, he seemed very far away. And as when a bearded lion, a ravening lion on the hills, hears the bleating of a fawn afar off, and rushes **forth** from his lair to seize it, his readiest meal, even so the mighty Heracles, in longing for the lad, sped through the trackless briars, and ranged over much country.

Reckless are lovers : great toils did Heracles bear, in hills and thickets wandering, and Iason's quest was all postponed to this. Now the ship abode with her tackling aloft, and the company gathered there,[1] but at midnight the young men were lowering the sails again, awaiting Heracles. But he wheresoever his feet might lead him went wandering in his fury, for the cruel Goddess of love was rending his heart within him.

Thus loveliest Hylas is numbered with the Blessed, but for a runaway they girded at Heracles, the heroes, because he roamed from Argo of the sixty oarsmen. **But** on foot he came to Colchis and inhospitable Phasis.

---

[1] Cf. Wordsworth's proposed conjecture—

μετάρσι᾽, ἐτῶν παρεόντων.

Meineke observes 'tota haec carminis pars luxata et foedissime depravata est.' There seems to be a rude early **pun** in lines 73, 74.

IDYL XIV

*This Idyl, like the next, is dramatic in form. One
Aeschines tells Thyonichus the story of his quarrel
with his mistress Cynisca. He speaks of taking
foreign service, and Thyonichus recommends that of
Ptolemy. The idyl was probably written at Alexandria, as a compliment to Ptolemy, and an induce-
ment to Greeks to join his forces. There is nothing,
however, to fix the date.*

Aeschines. All hail to the stout Thyonichus !

Thyonichus. As much to you, Aeschines.

Aeschines. How long it is since we met!

Thyonichus. Is it so long ? But why, pray,
this melancholy ?

Aeschines. I am not in the best of luck,
Thyonichus.

Thyonichus. 'Tis for that, then, you are so
lean, and hence comes this long moustache,
and these love-locks all adust. Just such a
figure was a Pythagorean that came here of
late, barefoot and wan,—and said he was an
Athenian. Marry, he too was in love, methinks,
with a plate of pancakes.

Aeschines. Friend, you will always have your

jest,—but beautiful Cynisca,—she flouts me!
I shall go mad some day, when no man looks
for it; I am but a hair's-breadth on the hither
side, even now.

Thyonichus. You are ever like this, dear
Aeschines, now mad, now sad, and crying for
all things at your whim. Yet, tell me, what is
your new trouble?

Aeschines. The Argive, and I, and the
Thessalian rough rider, Apis, and Cleunichus
the free lance, were drinking together, at
my farm. I had killed two chickens, and a
sucking pig, and had opened the Bibline wine
for them,—nearly four years old,—but fragrant
as when it left the wine-press. Truffles and
shellfish had been brought out, it was a
jolly drinking match. And when things were
now getting forwarder, we determined that each
of us should toast whom he pleased, in unmixed wine, only he must name his toast. So
we all drank, and called our toasts as had been
agreed. Yet She said nothing, though I was
there; how think you I liked that? 'Won't
you call a toast? You have seen the wolf!'
some one said in jest, 'as the proverb goes,'[1]
then she kindled; yes, you could easily have

---

[1] The reading—

οὐ φθεγξῇ; λύκον εἶδες; ἔπαιξέ τις, ὡς σοφός, εἶπε,—

makes good sense. ὡς σοφός is put in the mouth of the
girl, and would mean 'a good guess'! The allusion
of a guest to the superstition that the wolf struck people
dumb is taken by Cynisca for a reference to young
Wolf, her secret lover.

lighted a lamp at her face. There is one Wolf,
one Wolf there is, the son of Labes our neighbour,—he is tall, smooth-skinned, many think
him handsome. His was that illustrious love
in which she was pining, yes, and a breath
about the business once came secretly to my
ears, but I never looked into it, beshrew my
beard !

Already, mark you, we four men were deep
in our cups, when the Larissa man out of mere
mischief, struck up, ‘My Wolf,’ some Thessalian catch, from the very beginning. Then
Cynisca suddenly broke out weeping more bitterly than a six-year-old maid, that longs for
her mother’s lap. Then I,—you know me,
Thyonichus,—struck her on the cheek with
clenched fist,—one two ! She caught up her
robes, and forth she rushed, quicker than she
came. ‘Ah, my undoing’ (cried I), ‘I am
not good enough for you, then—you have a
dearer playfellow? well, be off and cherish
your other lover, ’tis for him your tears run big
as apples !’[1]

And as the swallow flies swiftly back to
gather a morsel, fresh food, for her young ones
under the eaves, still swifter sped she from her
soft chair, straight through the vestibule and
folding-doors, wherever her feet carried her.
So, sure, the old proverb says, ‘the bull has
sought the wild wood.’

Since then there are twenty days, and eight

[1] Or, as Wordsworth suggests, reading δάκρυσι,
‘ for him your cheeks are wet with tears.’

to these, and nine again, then ten others, to-
day is the eleventh, add two more, and it is
two months since we parted, and I have not
shaved, not even in Thracian fashion.[1]

And now Wolf is everything with her.   Wolf
finds the door open o' nights, and I am of no
account, not in the reckoning, like the wretched
men of Megara, in the place dishonourable.[2]

And if I could cease to love, the world would
wag as well as may be.   But now,—now,—as
they say, Thyonichus, I am like the mouse
that has tasted pitch.   And what remedy there
may be for a bootless love, I know not ; except
that Simus, he who was in love with the
daughter of Epicalchus, went over seas, and
came back heart-whole,—a man of my own
age.   And I too will cross the water, and prove
not the first, maybe, nor the last, perhaps, but
a fair soldier as times go.

Thyonichus. Would that things had gone to
your mind, Aeschines.   But if, in good earnest,
you are thus set on going into exile, PTOLEMY
is the free man's best paymaster !

Aeschines. And in other respects, what kind
of man ?

[1] Shaving in the bronze, and still more, of course, in
the stone age, was an uncomfortable and difficult pro-
cess.   The backward and barbarous Thracians were
therefore trimmed in the roughest way, like Aeschines,
with his long gnawed moustache.

[2] The Megarians having inquired of the Delphic
oracle as to their rank among Greek cities, were told
that they were absolute last, and not in the reckoning at
all.

Thyonichus. The free man's best paymaster! Indulgent too, the Muses' darling, a true lover, the top of good company, knows his friends, and still better knows his **enemies.** A great giver to many, refuses nothing that he is asked which to give may beseem a king, but, Aeschines, we should not always be asking. Thus, if you **are minded** to pin up **the top corner of** your cloak **over the** right shoulder, **and** if you have the heart to stand steady on both feet, and bide the brunt of a hardy targeteer, off instantly to Egypt! From the temples down-**ward we** all wax grey, and on to the chin creeps the rime of age, men must do somewhat while their knees are yet nimble.

*This famous idyl should rather, perhaps, be called a
  mimus. It describes the visit paid by two Syracusan
  women residing in Alexandria, to the festival of the
  resurrection of Adonis. The festival is given by
  Arsinoë, wife and sister of Ptolemy Philadelphus,
  and the poem cannot have been written earlier than
  his marriage, in 266 B.C. [?] Nothing can be more
  gay and natural than the **chatter** of **the women,
  which has changed no more** in two thousand years
  than the song of birds. Theocritus is believed to
  have had a model for this idyl in the Isthmiazusae
  of Sophron, an older poet. In the Isthmiazusae two
  ladies described the spectacle of the Isthmian games.*

Gorgo. Is Praxinoë at home ?

Praxinoë. Dear Gorgo, how long it is since
you have been here ! She *is* at home. **The**
wonder is **that you** have **got here at last !**
Eunoë, **see that she has a chair.** Throw a
cushion on **it too.**

Gorgo. It does **most charmingly** as it is.

Praxinoë. Do **sit down.**

Gorgo. Oh, **what** a thing spirit **is ! I have**
scarcely got to you alive, Praxinoë ! What
a huge crowd, what hosts of four-in-hands !
Everywhere cavalry boots, everywhere men **in**

uniform ! And the road is endless : yes, you really live *too* far away !

Praxinoë. It is all the fault of that madman of mine. Here he came to the ends of the earth and took—a hole, not a house, and all that we might not be neighbours. The jealous wretch, always the same, ever for spite !

Gorgo. Don't talk of your husband, Dinon, like that, my dear girl, before the little boy,— look how he is staring at you ! Never mind, Zopyrion, sweet child, she is not speaking about papa.

Praxinoë. Our Lady ! the child takes notice.[1]

Gorgo. Nice papa !

Praxinoë. That papa of his the other day— we call every day ' the other day '—went to get soap and rouge at the shop, and back he came to me with salt—the great big endless fellow !

Gorgo. Mine has the same trick, too, a perfect spendthrift—Diocleides ! Yesterday he got what he meant for five fleeces, and paid seven shillings a piece for—what do you suppose?—dogskins, shreds of old leather wallets, mere trash—trouble on trouble. But come, take your cloak and shawl. Let us be off to the palace of rich Ptolemy, the King, to see

---

[1] Our Lady, here, is Persephone. The ejaculation served for the old as well as for the new religion of Sicily. The dialogue is here arranged as in Fritzsche's text, and in line 8 his punctuation is followed.

the Adonis ; I hear the Queen has provided
something splendid !

Praxinoë. Fine folks do everything finely.

Gorgo. What a tale you will have to tell
about the things you have seen, to any one who
has not seen them ! It seems nearly time to
go.

Praxinoë. Idlers have always holiday. Eunoë, bring the water and put it down in the
middle of the room, lazy creature that you are.
Cats like always to sleep soft ! [1] Come, bustle,
bring the **water** ; quicker. I want water first,
and how she carries it ! give it me all the
same ; don't pour out so much, you extravagant
thing. Stupid girl ! Why are you wetting
my dress ? There, stop, I have washed my
hands, as heaven would have it. Where is the
key of the big chest ? Bring it here.

Gorgo. Praxinoë, that full body becomes you
wonderfully. Tell me how much did the stuff
cost you just off the **loom** ?

Praxinoë. Don't speak of it, Gorgo ! More
than eight pounds in good silver money,—and
the work on it ! I nearly slaved my soul out
over it !

Gorgo. Well, it is *most* successful ; all you
could wish.[2]

Praxinoë. Thanks for the pretty speech !

[1] If cats are meant, the proverb is probably Alexandrian. Common as cats were in Egypt, they were
late comers in Greece.

[2] Most of the dialogue has been distributed as in the
text of Fritzsche.

Bring my shawl, and set my hat on my head, the fashionable way. No, child, I don't mean to take you. Boo! Bogies! There's a horse that bites! Cry as much as you please, but I cannot have you lamed. Let us be moving. Phrygia take the child, and keep him amused, call in the dog, and shut the street door.

[*They go into the street.*

Ye gods, what a crowd! How on earth are we ever to get through this coil? They are like ants that no one can measure or number. Many a good deed have you done, Ptolemy; since your father joined the immortals, there's never a malefactor to spoil the passer - by, creeping on him in Egyptian fashion — oh! the tricks those perfect rascals used to play. Birds of a feather, ill jesters, scoundrels all! Dear Gorgo, what will become of us? Here come the King's war-horses! My dear man, don't trample on me. Look, the bay's rearing, see, what temper! Eunoë, you foolhardy girl, will you never keep out of the way? The beast will kill the man that's leading him. What a good thing it is for me that my brat stays safe at home.

Gorgo. Courage, Praxinoë. We are safe behind them, now, and they have gone to their station.

Praxinoë. There! I begin to be myself again. Ever since I was a child I have feared nothing so much as horses and the chilly snake. Come along, the huge mob is overflowing us.

Gorgo (*to an old Woman*). Are you from the Court, mother ?

Old Woman. I am, my child.

Praxinoë. Is it easy to get there ?

Old Woman. The Achaeans got into Troy by trying, my prettiest of ladies. Trying will do everything in the long run.

Gorgo. The old wife has spoken her oracles, and off she goes.

Praxinoë. Women know everything, yes, and how Zeus married Hera !

Gorgo. See Praxinoë, what a crowd there is about the doors.

Praxinoë. Monstrous, Gorgo ! Give me your hand, and you, Eunoë, catch hold of Eutychis ; never lose hold of her, for fear lest you get lost. Let us all go in together ; Eunoë, clutch tight to me. Oh, how tiresome, Gorgo, my muslin veil is torn in two already ! For heaven's sake, sir, if you ever wish to be fortunate, take care of my shawl !

Stranger. I can hardly help myself, but for all that I will be as careful as I can.

Praxinoë. How close - packed the mob is, they hustle like a herd of swine.

Stranger. Courage, lady, all is well with us now.

Praxinoë. Both this year and for ever may all be well with you, my dear sir, for your care of us. A good kind man ! We're letting Eunoë get squeezed — come, wretched girl, push your way through. That is the way. We are all on the right side of the door, quoth

the bridegroom, when he had shut himself in
with his bride.

Gorgo. Do come here, Praxinoë. Look first
at these embroideries. How light and how
lovely! You will call them the garments of
the gods.

Praxinoë. Lady Athene, what spinning
women wrought them, what painters designed
these drawings, so true they are? How
naturally they stand and move, like living
creatures, not patterns woven. What a clever
thing is man! Ah, and himself—Adonis—
how beautiful to behold he lies on his silver
couch, with the first down on his cheeks, the
thrice-beloved Adonis,—Adonis beloved even
among the dead.

A Stranger. You weariful women, do cease
your endless cooing talk! They bore one to
death with their eternal broad vowels!

Gorgo. Indeed! And where may this
person come from? What is it to you if we
are chatterboxes! Give orders to your own
servants, sir. Do you pretend to command
ladies of Syracuse? If you must know, we
are Corinthians by descent, like Bellerophon
himself, and we speak Peloponnesian. Dorian
women may lawfully speak Doric, I presume?

Praxinoë. Lady Persephone, never may we
have more than one master. I am not afraid
of *your* putting me on short commons.

Gorgo. Hush, hush, Praxinoë—the Argive
woman's daughter, the great singer, is beginning the *Adonis;* she that won the prize last

G

year for dirge-singing.[1]   I am sure she will
give us something lovely ; see, she is preluding
with her airs and graces.

The Psalm of Adonis.

O Queen that lovest Golgi, and Idalium, and
the steep of Eryx, O Aphrodite, that playest
with gold, lo, from the stream eternal of
Acheron they have brought back to thee
Adonis—even in the twelfth month they have
brought him, the dainty-footed Hours. Tardi-
est of the Immortals are the beloved Hours,
but dear and desired they come, for always, to
all mortals, they bring some gift with them.
O Cypris, daughter of Diônê, from mortal to
immortal, so men tell, thou hast changed
Berenice, dropping softly in the woman's breast
the stuff of immortality.

Therefore, for thy delight, O thou of many
names and many temples, doth the daughter
of Berenice, even Arsinoë, lovely as Helen,
cherish Adonis with all things beautiful.

Before him lie all ripe fruits that the tall
trees' branches bear, and the delicate gardens,
arrayed in baskets of silver, and the golden
vessels are full of incense of Syria. And all
the dainty cakes that women fashion in the
kneading-tray, mingling blossoms manifold with
the white wheaten flour, all that is wrought of
honey sweet, and in soft olive oil, all cakes
fashioned in the semblance of things that fly,

---

[1] Reading πέρυσιν.

and of things that creep, lo, here they are set
before him.

Here are built for him shadowy bowers of
green, all laden with tender anise, and children
flit overhead—the little Loves—as the young
nightingales perched upon the trees fly forth
and try their wings from bough to bough.

O the ebony, O the gold, O the twin eagles
of white ivory that carry to Zeus the son of
Cronos his darling, his cup-bearer! O the
purple coverlet strewn above, more soft than
sleep! So Miletus will say, and whoso feeds
sheep in Samos.

Another bed is strewn for beautiful Adonis,
one bed Cypris keeps, and one the rosy-armed
Adonis. A bridegroom of eighteen or nine-
teen years is he, his kisses are not rough, the
golden down being yet upon his lips! And
now, good-night to Cypris, in the arms of her
lover! But lo, in the morning we will all of
us gather with the dew, and carry him forth
among the waves that break upon the beach,
and with locks unloosed, and ungirt raiment
falling to the ankles, and bosoms bare will we
begin our shrill sweet song.

Thou only, dear Adonis, so men tell, thou
only of the demigods dost visit both this world
and the stream of Acheron. For Agamemnon
had no such lot, nor Aias, that mighty lord of
the terrible anger, nor Hector, the eldest born
of the twenty sons of Hecabe, nor Patroclus,
nor Pyrrhus, that returned out of Troyland,
nor the heroes of yet more ancient days, the

Lapithae and Deucalion's sons, nor the sons of
Pelops, and the chiefs of Pelasgian Argos.
Be gracious now, dear Adonis, and propitious
even in the coming year. Dear to us has
thine advent been, Adonis, and dear shall it
be when thou comest again.

Gorgo. Praxinoë, the woman is cleverer than
we fancied! Happy woman to know so much,
thrice happy to have so sweet a voice. Well,
all the same, it is time to be making for home.
Diocleides has not had his dinner, and the
man is all vinegar,—don't venture near him
when he is kept waiting for dinner. Farewell,
beloved Adonis, may you find us glad at your
next coming!

IDYL XVI

*In 265 B.C. Sicily was devastated by the Carthaginians,
and by the companies of disciplined free-lances who
called themselves* Mamertines, *or* Mars's men. *The
hopes of the Greek inhabitants of the island were
centred in Hiero, son of Hierocles, who was about
to besiege Messana (then held by the Carthaginians)
and who had revived the courage of the Syracusans.
To him Theocritus addressed this idyl, in which he
complains of the sordid indifference of the rich,
rehearses the merits of song, dilates on the true
nature of wealth, and of the happy life, and finally
expresses his hope that Hiero will rid the isle of the
foreign foe, and will restore peace and pastoral joys.
The idyl contains some allusions to Simonides, the
old lyric poet, and to his relations with the famous
Hiero tyrant of Syracuse.*

EVER is this the care of the maidens of
Zeus, ever the care of minstrels, to sing the
Immortals, to sing the praises of noble men.
The Muses, lo, are Goddesses, of Gods the
Goddesses sing, but we on earth are mortal
men ; let us mortals sing of mortals. Ah, who
of all them that dwell beneath the grey morning,
will open his door and gladly receive our Graces
within his house ? who is there that will not
send them back again without a gift? And

they with looks askance, and naked feet come
homewards, and sorely they upbraid me when
they have gone on a vain journey, and listless
again in the bottom of their empty coffer, they
dwell with heads bowed over **their** chilly knees,
where is their drear abode, **when gainless**
they return.

Where is there such an one, among men to-
day? Where is he that will befriend him that
speaks his praises? I know not, for now no
longer, as of old, are men eager to win the
renown of noble deeds, nay, they are the slaves
of gain! Each man clasps his hands below
the purse-fold of his gown, and looks about to
spy whence he may get him money: the very
rust is too precious to be rubbed off for a gift.
Nay, each has his ready saw; *the shin is
further than the knee ; first let me get my own !
'Tis the Gods' affair to honour minstrels !
Homer is enough for every one, who wants to
hear any other ? He is the best of bards who
takes nothing that is mine.*

O foolish **men, in the** store of gold **un-**
counted, what gain have ye? Not in this do
the wise find the true enjoyment of wealth, but
in that they can indulge their own desires, and
something bestow on one of the minstrels, and
do good deeds to many of their kin, and to
many another man ; and always give altar-rites
to the Gods, nor ever play the churlish host,
but kindly entreat the guest at table, and
speed him when he would be gone. And this,
above all, to honour the holy interpreters of the

Muses, that so thou mayest have a goodly
fame, even when hidden in Hades, nor ever
moan without renown by the chill water of
Acheron, like one whose palms the spade has
hardened, some landless man bewailing the
poverty that is all his heritage.

Many were the thralls that in the palace of
Antiochus, and of king Aleuas drew out their
monthly dole, many the calves that were driven
to the penns of the Scopiadae, and lowed with
the horned kine : countless on the Crannonian
plain did shepherds pasture beneath the sky
the choicest sheep of the hospitable Creondae,
yet from all this they had no joy, when once
into the wide raft of hateful Acheron they had
breathed sweet life away ! Yea, unremembered
(though they had left all that rich store), for
ages long would they have lain among the dead
forlorn, if a name among later men the skilled
Ceian minstrel had spared to bestow, singing
his bright songs to a harp of many strings.
Honour too was won by the swift steeds that
came home to them crowned from the sacred
contests.

And who would ever have known the Lycian
champions of time past, who Priam's long-
haired sons, and Cycnus, white of skin as a
maiden, if minstrels had not chanted of the war
cries of the old heroes ? Nor would Odysseus
have won his lasting glory, for all his ten years'
wandering among all folks ; and despite the
visit he paid, he a living man, to inmost Hades,
and for all his escape from the murderous

Cyclops's cave,—unheard too were the names
of the swineherd Eumaeus, and of Philoetius,
busy with the kine of the herds ; yea, and even
of Laertes, high of heart ; if the songs of the
Ionian man had not kept them in renown.

From the Muses comes a goodly report to
men, but the living heirs devour the possessions
of the dead.   But, lo, it is as light labour to
count the waves upon the beach, as many as
wind and grey sea-tide roll upon the shore, or
in violet-hued water to cleanse away the stain
from a potsherd, as to win favour from a man
that is smitten with the greed of gain.   Good-
day to such an one, and countless be his coin,
and ever may he be possessed by a longing
desire for more !   But I for my part would
choose honour and the loving-kindness of men,
far before wealth in mules and horses.

I am seeking to what mortal I may come, a
welcome guest, with the help of the Muses,
for hard indeed do minstrels find the ways,
who go uncompanioned by the daughters of
deep-counselling Zeus.   Not yet is the heaven
aweary of rolling the months onwards, and the
years, and many a horse shall yet whirl the
chariot wheels, and the man shall yet be found,
who will take me for his minstrel ; a man of
deeds like those that great Achilles wrought, or
puissant Aias, in the plain of Simois, where is
the tomb of Phrygian Ilus.

Even now the Phoenicians that dwell beneath the setting sun on the spur of Libya,
shudder for dread, even now the Syracusans

poise lances in rest, and their arms are burdened by the linden shields. Among them
Hiero, like the mighty men of old, girds himself for fight, and the horse - hair crest is
shadowing his helmet. Ah, Zeus, our father
renowned, and ah, lady Athene, and O thou
Maiden that with the Mother dost possess the
great burg of the rich Ephyreans, by the
water of Lusimeleia,[1] would that dire necessity
may drive our foemen from the isle, along the
Sardinian wave, to tell the doom of their friends
to children and to wives—messengers easy to
number out of so many warriors ! But as for
our cities may they again be held by their
ancient masters,—all the cities that hostile
hands have utterly spoiled. May our people
till the flowering fields, and may thousands of
sheep unnumbered fatten 'mid the herbage, and
bleat along the plain, while the kine as they
come in droves to the stalls warn the belated
traveller to hasten on his way. May the fallows
be broken for the seed-time, while the cicala,
watching the shepherds as they toil in the sun,
in the shade of the, trees doth sing on the top-
most sprays. May spiders weave their delicate
webs over martial gear, may none any more so
much as name the cry of onset !

But the fame of Hiero may minstrels bear
aloft, across the Scythian sea, and where
Semiramis reigned, that built the mighty wall,

---

[1] *I.e.* Syracuse, a colony of the Ephyraeans or Corinthians.   The Maiden is Persephone, the Mother
Demeter.

and made it fast with slime for mortar. I am
but one of many that are loved by the daughters
of Zeus, and they all are fain to sing of Sicilian
Arethusa, with the people of the isle, and the
warrior Hiero. O Graces, ye Goddesses,
adored of Eteocles, ye that love Orchomenos
of the Minyae, the ancient enemy of Thebes,
when no man bids me, let me abide at home,
but to the houses of such as bid me, boldly let
me come with my Muses. Nay, neither the
Muses nor you Graces will I leave behind, for
without the Graces what have men that is
desirable? with the Graces of song may I
dwell for ever!

The poet praises Ptolemy Philadelphus in a strain of almost religious adoration. Hauler, in his Life of Theocritus, dates the poem about 259 B.C., but it may have been many years earlier.

FROM Zeus let us begin, and with Zeus make end, ye Muses, whensoever we chant in songs the chiefest of immortals! But of men, again, let Ptolemy be named, among the foremost, and last, and in the midmost place, for of men he hath the pre-eminence. The heroes that in old **days** were begotten of the demigods, wrought noble deeds, and chanced on minstrels skilled, but I, with what skill I have in song, would fain make my hymn of Ptolemy, and hymns are the glorious meed, yea, of the very immortals.

When the feller hath come up to wooded Ida, he glances around, so many are the trees, to **see** whence he should begin his labour. Where first shall *I* begin the tale, for there are countless things ready for the telling, wherewith the Gods have graced the most excellent of kings?

Even by virtue of his sires, how mighty was he to accomplish some great work,—Ptolemy

son of Lagus,—when he had stored in his
mind such a design, as no other man was able
even to devise ! Him hath the Father stablished
in the same honour as the blessed immortals,
and for him a golden mansion in the house
of Zeus is builded ; beside him is throned
Alexander, that dearly loves him, Alexander, a
grievous god to the white-turbaned Persians.

And over against them is set the throne of
Heracles, the slayer of the Bull, wrought of
stubborn adamant. There holds he festival
with the rest of the heavenly host, rejoicing
exceedingly in his far-off children's children, for
that the son of Cronos hath taken old age
clean away from their limbs, and they are
called immortals, being his offspring. For the
strong son of Heracles is ancestor of the twain,
and both are reckoned to Heracles, on the
utmost of the lineage.

Therefore when he hath now had his fill of
fragrant nectar, and is going from the feast to
the bower of his bed-fellow dear, to one of his
children he gives his bow, and the quiver that
swings beneath his elbow, to the other his
knotted mace of iron. Then they to the
ambrosial bower of white-ankled Hera, convey
the weapons and the bearded son of Zeus.

Again, how shone renowned Berenice among
the wise of womankind, how great a boon was
she to them that begat her ! Yea, in her
fragrant breast did the Lady of Cyprus, the
queenly daughter of Dione, lay her slender
hands, wherefore they say that never any

woman brought man such delight as came
from the love borne to his wife by Ptolemy.
And verily he was loved again with far greater
love, and in such a wedlock a man may well
trust all his house to his children, whensoever he
goes to the bed of one that loves him as he loves
her. But the mind of a woman that loves not is
set ever on a stranger, and she hath children
at her desire, but they are never like the father.

O thou that amongst the Goddesses hast
the prize of beauty, O Lady Aphrodite, thy
care was she, and by thy favour the lovely
Berenice crossed not Acheron, the river of
mourning, but thou didst catch her away, ere
she came to the dark water, and to the still-
detested ferryman of souls outworn, and in thy
temple didst thou instal her, and gavest her a
share of thy worship. Kindly is she to all
mortals, and she breathes into them soft desires,
and she lightens the cares of him that is in
longing.

O dark-browed lady of Argos,[1] in wedlock
with Tydeus didst thou bear slaying Diomede,
a hero of Calydon, and, again, deep-bosomed
Thetis to Peleus, son of Aeacus, bare the spearman Achilles. But thee, O warrior Ptolemy, to
Ptolemy the warrior bare the glorious Berenice !
And Cos did foster thee, when thou wert still
a child new-born, and received thee at thy
mother's hand, when thou saw'st thy first
dawning. For there she called aloud on
Eilithyia, loosener of the girdle ; she called,

[1] Deipyle, daughter of Adrastus.

the daughter of Antigone, when heavy on her
came the pangs of childbirth. And Eilithyia
was present to help her, and so poured over all
her limbs release from pain. Then the beloved child was born, his father's very counter-
part. And Cos brake forth into a cry, when
she beheld it, and touching the child with kind
hands, she said :

'Blessed, O child, mayst thou be, and me
mayst thou honour even as Phoebus Apollo
honours Delos of the azure crown, yea, stablish
in the same renown the Triopean hill, and allot
such glory to the Dorians dwelling nigh, as that
wherewithal Prince Apollo favours Rhenaea.'

Lo, thus spake the Isle, but far aloft under
the clouds a great eagle screamed thrice aloud,
the ominous bird of Zeus. This sign, methinks,
was of Zeus ; Zeus, the son of Cronos, in his
care hath awful kings, but he is above all,
whom Zeus loved from the first, even from his
birth. Great fortune goes with him, and much
land he rules, and wide sea.

Countless are the lands, and tribes of men
innumerable win increase of the soil that waxeth
under the rain of Zeus, but no land brings
forth so much as low-lying Egypt, when Nile
wells up and breaks the sodden soil. Nor is
there any land that hath so many towns of men
skilled in handiwork ; therein are three cen-
turies of cities builded, and thousands three,
and to these three myriads, and cities twice
three, and beside these, three times nine, and
over them all high-hearted Ptolemy is king.

Yea, and he taketh him a portion of Phoenicia, and of Arabia, and of Syria, and of Libya,
and the black Aethiopians. And he is lord of
all the Pamphylians, and the Cilician warriors,
and the Lycians, and the Carians, that joy in
battle, **and** lord of the **isles of the** Cyclades,—
since his are the best **of** ships that sail over
the deep,—yea, all the sea, and land and **the**
sounding rivers are ruled by Ptolemy. Many
are his horsemen, and many his targeteers that
go clanging in harness of shining bronze. And
in weight of wealth he surpasses all kings;
such treasure comes day by day from every
side to his rich palace, while the people are
busy about their labours in peace. For never
hath a foeman marched up the bank of teaming
Nile, and raised the cry of war in villages not
his own, nor hath any cuirassed enemy leaped
ashore from his swift ship, to harry the kine **of**
Egypt. So mighty **a** hero hath his throne
established in the broad plains, even Ptolemy
of the fair hair, a spearman skilled, whose care
is above all, as a good king's should be, to
keep all the heritage of his fathers, and yet
more he himself doth win. Nay, nor useless
in *his* wealthy house, is the gold, like piled
stores of the still toilsome ants, but the glorious
temples of the gods have their rich share, for
constant first-fruits he renders, with **many an-**
other due, and much is lavished on mighty
kings, much on cities, much on faithful friends.
And never to the sacred contests of Dionysus
comes any man that is skilled to raise the shrill

sweet song, but Ptolemy gives him a guerdon worthy of his art. And the interpreters of the Muses sing of Ptolemy, in return for his favours. Nay, what fairer thing might befall a wealthy man, than to win a goodly renown among mortals?

This abides even by the sons of Atreus, but all those countless treasures that they won, when they took the mighty house of Priam, are hidden away in the mist, whence there is no returning.

Ptolemy alone presses his own feet in the footmarks, yet glowing in the dust, of his fathers that were before him. To his mother dear, and his father he hath stablished fragrant temples ; therein has he set their images, splendid with gold and ivory, to succour all earthly men. And many fat thighs of kine doth he burn on the empurpled altars, as the months roll by, he and his stately wife ; no nobler lady did ever embrace a bridegroom in the halls, who loves, with her whole heart, her brother, her lord. On this wise was the holy bridal of the Immortals, too, accomplished, even of the pair that great Rhea bore, the rulers of Olympus ; and one bed for the slumber of Zeus and of Hera doth Iris strew, with myrrh-anointed hands, the virgin Iris.

Prince Ptolemy, farewell, and of thee will I make mention, even as of the other demigods ; and a word methinks I will utter not to be rejected of men yet unborn,—excellence, how-beit, thou shalt gain from Zeus.

IDYL XVIII

This epithalamium may have been written for the wedding of a friend of the poet's. The idea is said to have been borrowed from an old poem by Stesichorus. The epithalamium was chanted at night by a chorus of girls, outside the bridal chamber. Compare the conclusion of the hymn of Adonis, in the fifteenth Idyl.

IN Sparta, once, to the house of fair-haired Menelaus, **came** maidens with **the** blooming hyacinth in their hair, and before the new painted chamber arrayed their dance,—twelve maidens, the first in the city, the glory of Laconian girls,—what time the younger Atrides had wooed and won Helen, and closed the door of the bridal-bower on the beloved daughter of Tyndarus. Then sang they all in harmony, beating time with woven paces, and the house rang round with the bridal song.

The Chorus.

Thus early **art** thou sleeping, dear bridegroom, say are thy limbs heavy with slumber, or art thou all too fond of sleep, or hadst thou perchance drunken over well, ere thou didst

H

fling thee to thy rest? Thou shouldst have slept betimes, and alone, if thou wert so fain of sleep; thou shouldst have left the maiden with maidens beside her mother dear, to play till deep in the dawn, for to-morrow, and next day, and for all the years, Menelaus, she is thy bride.

O happy bridegroom, some good spirit sneezed out on thee a blessing, as thou wert approaching Sparta whither went the other princes, that so thou mightst win thy desire! Alone among the demigods shalt thou have Zeus for father! Yea, and the daughter of Zeus has come beneath one coverlet with thee, so fair a lady, peerless among all Achaean women that walk the earth. Surely a wondrous child would she bear thee, if she bore one like the mother!

For lo, we maidens are all of like age with her, and one course we were wont to run, anointed in manly fashion, by the baths of Eurotas. Four times sixty girls were we, the maiden flower of the land, but of us all not one was faultless, when matched with Helen.

As the rising Dawn shows forth her fairer face than thine, O Night, or as the bright Spring, when Winter relaxes his hold, even so amongst us still she shone, the golden Helen. Even as the crops spring up, the glory of the rich plough land;[1] or, as is the cypress in the garden; or, in a chariot, a horse of Thessalian

---

[1] Reading—πιείρᾳ ἅτε λῇον ἀνέδραμε κόσμος ἀρούρᾳ. See also Wordsworth's note on line 26.

breed, even so is rose-red Helen the glory of
Lacedaemon. No other in her basket of wool
winds forth such goodly work, and none cuts
out, from between the mighty beams, a closer
warp than that her shuttle weaves in the carven
loom. Yea, and of a truth none other smites
the lyre, hymning Artemis and broad-breasted
Athene, with such skill as Helen, within whose
eyes dwell all the Loves.

O fair, O gracious damsel, even now art
thou a wedded wife ; but we will go forth right
early to the course we ran, and to the grassy
meadows, to gather sweet-breathing coronals of
flowers, thinking often upon thee, Helen, even
as youngling lambs that miss the teats of the
mother-ewe. For thee first will we twine a
wreath of lotus flowers that lowly grow, and
hang it on a shadowy plane tree, for thee first
will we take soft oil from the silver phial, and
drop it beneath a shadowy plane tree, and
letters will we grave on the bark, in Dorian
wise, so that the wayfarer may read :

WORSHIP ME, I AM THE TREE OF HELEN.

Good night, thou bride, good night, thou
groom that hast won a mighty sire ! May
Leto, Leto, the nurse of noble offspring, give
you the blessing of children ; and may Cypris,
divine Cypris, grant you equal love, to cherish
each the other ; and may Zeus, even Zeus the
son of Cronos, give you wealth imperishable,
to be handed down from generation to generation of the princes.

Sleep ye, breathing love and desire each
into the other's breast, but forget not to wake
in the dawning, and at dawn we too will come,
when the earliest cock shrills from his perch,
and raises his feathered neck.

*Hymen, O Hymenae, rejoice thou in this
bridal.*

IDYL XIX

This little piece is but doubtfully ascribed to Theocritus. The motif is that of a well-known Anacreontic Ode. The idyl has been translated by Ronsard.

THE thievish Love,—a cruel bee once stung him, as he was rifling honey from the hives, and pricked his finger-tips all; then he was in pain, and blew upon his hand, and leaped, and stamped the ground. And then he showed his hurt to Aphrodite, and made much complaint, how that the bee is a tiny creature, and yet what wounds it deals! And his mother laughed out, and said, 'Art thou not even such a creature as the bees, for tiny art thou, but what wounds thou dealest!'

IDYL XX

A herdsman, who had been contemptuously rejected by Eunica, a girl of the town, protests that he is beautiful, and that Eunica is prouder than Cybele, Selene, and Aphrodite, all of whom loved mortal herdsmen. For grammatical and other reasons, some critics consider this idyl apocryphal.

EUNICA laughed out at me when sweetly I would have kissed her, and taunting me, thus she spoke : 'Get thee gone from me ! Wouldst thou kiss me, wretch ; thou—a neatherd ? I never learned to kiss in country fashion, but to press lips with city gentlefolks. Never hope to kiss my lovely mouth, nay, not even in a dream. How thou dost look, what chatter is thine, how countrified thy tricks are, how delicate thy talk, how easy thy tattle ! And then thy beard—so soft ! thy elegant hair ! Why, thy lips are like some sick man's, thy hands are black, and thou art of evil savour. Away with thee, lest thy presence soil me !' These taunts she mouthed, and thrice spat in the breast of her gown, and stared at me all over from head to feet ; shooting out her lips, and glancing with half-shut eyes, writhing her beautiful body, and so

sneered, and laughed me to scorn. And instantly my blood boiled, and I grew red under
the sting, as a rose with dew. And she went
off and left me, but I bear angry pride deep
in my heart, that I, the handsome shepherd,
should have been mocked by a wretched light-
o'-love.

Shepherds, tell me the very truth ; am I not
beautiful? Has some God changed me suddenly to another man? Surely a sweet grace
ever blossomed round me, till this hour, like
ivy round a tree, and covered my chin, and
about my temples fell my locks, like curling
parsley-leaves, and white shone my forehead
above my dark eyebrows. Mine eyes were
brighter far than the glance of the grey-eyed
Athene, my mouth than even pressed milk was
sweeter, and from my lips my voice flowed
sweeter than honey from the honeycomb.
Sweet too, is my music, whether I make melody
on pipe, or discourse on the flute, or reed, or
flageolet. And all the mountain-maidens call
me beautiful, and they would kiss me, all of
them. But the city girl did not kiss me, but
ran past me, because I am a neatherd, and
she never heard how fair Dionysus in the dells
doth drive the calves, and knows not that
Cypris was wild with love for a herdsman, and
drove afield in the mountains of Phrygia ; ay,
and Adonis himself, — in the oakwood she
kissed, in the oakwood she bewailed him.
And what was Endymion? was he not a neatherd? whom nevertheless as he watched his

herds Selene saw and loved, and from Olympus descending she came to the Latmian glade, and lay in one couch with the boy ; and thou, Rhea, dost weep for thy herdsman.

And didst not thou, too, Son of Cronos, take the shape of a wandering bird, and all for a cowherd boy ?

But Eunica alone would not kiss the herdsman ; Eunica, she that is greater than Cybele, and Cypris, and Selene !

Well, Cypris, never mayst thou, in city or on hillside, kiss thy darling,[1] and lonely all the long night mayst thou sleep !

---

[1] For ἁδέα Wordsworth and Hermann conjecture ῎Αρεα. The sense would be that Eunica, who thinks herself another Cypris, or Aphrodite is, in turn, to be rejected by her Ares, her soldier-lover, as she has rejected the herdsman.

After some verses addressed to Diophantus, a friend about whom nothing is known, the poet describes the toilsome life of two old fishermen. One of them has dreamed of catching a golden fish, and has sworn, in his dream, never again to tempt the sea. The other reminds him that his oath is as empty as his vision, and that he must angle for common fish, if he would not starve among his golden dreams. The idyl is, unfortunately, corrupt beyond hope of certain correction.

'TIS Poverty alone, Diophantus, that awakens the arts ; Poverty, the very teacher of labour. Nay, not even sleep is permitted, by weary cares, to men that live by toil, and if, for a little while, one close his eyes [1] in the night, cares throng about him, and suddenly disquiet his slumber.

Two fishers, on a time, two old men, together lay and slept ; they had strown the dry sea-moss for a bed in their wattled cabin, and there they lay against the leafy wall. Beside them were strewn the instruments of their toilsome hands, the fishing-creels, the rods of reed, the hooks, the sails bedraggled with sea-

---

[1] Reading ἐπιμύσσῃσι.

spoil,[1] the lines, the weels, the lobster pots woven of rushes, the seines, two oars,[2] and an old coble upon props. Beneath their heads was a scanty matting, their clothes, their sailor's caps. Here was all their toil, here all their wealth. The threshold had never a door, nor a watch-dog;[3] all things, all, to them seemed superfluity, for Poverty was their sentinel. They had no neighbour by them, but ever against their narrow cabin gently floated up the sea.

The chariot of the moon had not yet reached the mid-point of her course, but their familiar toil awakened the fishermen; from their eyelids they cast out slumber, and roused their souls with speech.[4]

Asphalion. They lie all, my friend, who say that the nights wane short in summer, when Zeus brings the long days. Already have I seen ten thousand dreams, and the dawn is not yet. Am I wrong, what ails them, the nights are surely long?

The Friend. Asphalion, thou blamest the beautiful summer! It is not that the season hath wilfully passed his natural course, but care, breaking thy sleep, makes night seem long to thee.

Asphalion. Didst ever learn to interpret dreams? for good dreams have I beheld. I

[1] Reading τὰ φυκιοέντα τε λαίφη.
[2] κώπα.
[3] οὐδὸς δ᾽ οὐχὶ θύραν εἶχ᾽, and in the next line ἁ γὰρ πενία σφας ἐτήρει.　　　[4] αὐδάν.

would not have thee to go without thy share in
my vision ; even as we go shares in the fish we
catch, so share all my dreams ! Sure, thou art
not to be surpassed in wisdom ; and he is the
best interpreter of dreams that hath wisdom for
his teacher. Moreover, we have time to idle
in, for what could a man find to do, lying on a
leafy bed beside the wave and slumbering not ?
Nay, the ass is among the thorns, the lantern
in the town hall, for, they say, it is always
sleepless.[1]

The Friend. Tell me, then, the vision of the
night ; nay, tell all to thy friend.

Asphalion. As I was sleeping late, amid the
labours of the salt sea (and truly not too full-
fed, for we supped early if thou dost remember,
and did not overtax our bellies), I saw myself
busy on a rock, and there I sat and watched
the fishes, and kept spinning the bait with the
rods. And one of the fish nibbled, a fat one,
for in sleep dogs dream of bread, and of fish
dream I. Well, he was tightly hooked, and
the blood was running, and the rod I grasped
was bent with his struggle. So with both
hands I strained, and had a sore tussle for the
monster. How was I ever to land so big a

---

1 Reading, with Fritzsche—

ἀλλ᾽ ὄνος ἐν ῥάμνῳ, τό τε λύχνιον ἐν πρυτανείῳ
φαντὶ γὰρ ἀγρυπνίαν τόδ᾽ ἔχειν.

The lines seem to contain two popular saws, of which it
is difficult to guess the meaning. The first saw appears
to express helplessness ; the second, to hint that such
comforts as lamps lit all night long exist in towns, but
are out of the reach of poor fishermen.

fish with hooks all too slim ?   Then just to
remind him he was hooked, I gently pricked
him,[1] pricked, and slackened, and, as he did
not run, I took in line.   My toil was ended
with the sight of my prize ; I drew up a golden
fish, lo you, a fish all plated thick with gold !
Then fear took hold of me, lest he might be
some fish beloved of Posidon, or perchance
some jewel of the sea-grey Amphitrite.   Gently
I unhooked him, lest ever the hooks should
retain some of the gold of his mouth.   Then I
dragged him on shore with the ropes,[2] and
swore that never again would I set foot on
sea, but abide on land, and lord it over the
gold.

This was even what wakened me, but, for

[1] Reading ἠρέμ' ἔνυξα καὶ νύξας ἐχάλαξα. Asphalion
first hooked his fish, which ran gamely, and nearly
doubled up the rod.   Then the fish sulked, and the
angler half despaired of landing him.   To stir the sullen
fish, he 'reminded him of his wound,' probably, as we
do now, by keeping a tight line, and tapping the butt of
the rod.   Then he slackened, giving the fish line in case
of a sudden rush ; but as there was no such rush, he
took in line, or perhaps only showed his fish the butt
(for it is not probable that Asphalion had a reel), and
so landed him.   The Mediterranean fishers generally
toss the fish to land with no display of science, but
Asphalion's imaginary capture was a monster.

[2] It is difficult to understand this proceeding.   Perhaps Asphalion had some small net fastened with strings
to his boat, in which he towed fish to shore, that the
contact with the water might keep them fresher than
they were likely to be in the bottom of the coble.   On
the other hand, Asphalion was fishing from a rock.
His dream may have been confused.

the rest, set thy mind to it, my friend, for I am in dismay about the oath I swore.

The Friend. Nay, never fear, thou art no more sworn than thou hast found the golden fish of thy vision ; dreams are but lies. But if thou wilt search these waters, wide awake, and not asleep, there is some hope in thy slumbers ; seek the fish of flesh, lest thou die of famine with all thy dreams of gold !

IDYL XXII

THE DIOSCURI

*This is a hymn, in the Homeric manner, to Castor and
Polydeuces. Compare the life and truth of the
descriptions of nature, and of the boxing-match,
with the frigid manner of Apollonius Rhodius.—*
Argonautica, II. I. *seq.*

WE hymn the children twain of Leda, and of
aegis-bearing Zeus,—Castor, and Pollux, the
boxer dread, when he hath harnessed his
knuckles in thongs of ox-hide. Twice hymn
we, and thrice the stalwart sons of the daughter
of Thestias, the two brethren of Lacedaemon.
Succourers are they of men in the very thick of
peril, and of horses maddened in the bloody
press of battle, and of ships that, defying the
stars that set and rise in heaven, have en-
countered the perilous breath of storms. The
winds raise huge billows about their stern, yea,
or from the prow, or even as each wind wills,
and cast them into the hold of the ship, and
shatter both bulwarks, while with the sail hangs
all the gear confused and broken, and the
storm-rain falls from heaven as night creeps on,

and the wide sea rings, being lashed by the
gusts, and by showers of iron hail.

Yet even so do ye draw forth the ships from
the abyss, with their sailors that looked immediately to die ; and instantly the winds are
still, and there is an oily calm along the sea,
and the clouds flee apart, this way and that,
also the *Bears* appear, and in the midst, dimly
seen, the *Asses' manger*, declaring that all is
smooth for sailing.

O ye twain that aid all mortals, O beloved
pair, ye knights, ye harpers, ye wrestlers, ye
minstrels, of Castor, or of Polydeuces first shall
I begin to sing ? Of both of you will I make
my hymn, but first will I sing of Polydeuces.

Even already had Argo fled forth from the
Clashing Rocks, and the dread jaws of snowy
Pontus, and was come to the land of the
Bebryces, with her crew, dear children of the
gods.   There all the heroes disembarked, down
one ladder, from both sides of the ship of
Iason. When they had landed on the deep
seashore and a sea-bank sheltered from the
wind, they strewed their beds, and their hands
were busy with firewood.[1]

Then Castor of the swift steeds, and swart
Polydeuces, these twain went wandering alone,
apart from their fellows, and marvelling at
all the various wildwood on the mountain.
Beneath a smooth cliff they found an ever-
flowing spring filled with the purest water, and

---

[1] πυρεῖα appear to have been 'fire sticks,' by rub-
bing which together the heroes struck a light.

the pebbles below shone like crystal or silver
from the deep.   Tall fir trees grew thereby,
and white poplars, and planes, and cypresses
with their lofty tufts of leaves, and there
bloomed all fragrant flowers that fill the
meadows when early summer is waning—dear
work-steads of the hairy bees.   But there a
monstrous man was sitting in the sun, terrible
of aspect ; the bruisers' hard fists had crushed
his ears, and his mighty breast and his broad
back were domed with iron flesh, like some
huge statue of hammered iron.   The muscles
on his brawny arms, close by the shoulder,
stood out like rounded rocks, that the winter
torrent has rolled, and worn smooth, in the
great swirling stream, but about his back and
neck was draped a lion's skin, hung by the
claws.   Him first accosted the champion,
Polydeuces.

Polydeuces. Good luck to thee, stranger,
whosoe'er thou art !   What men are they that
possess this land ?

Amycus. What sort of luck, when I see men
that I never saw before ?

Polydeuces. Fear not !   Be sure that those
thou look'st on are neither evil, nor the children
of evil men.

Amycus. No fear have I, and it is not for
thee to teach me that lesson.

Polydeuces. Art thou a savage, resenting all
address, or some vainglorious man ?

Amycus. I am that thou see'st, and on thy
land, at least, I trespass not.

Polydeuces. Come, and with kindly gifts return homeward again !

Amycus. Gift me no gifts, none such have I ready for thee.

Polydeuces. Nay, wilt thou not even grant us leave to taste this spring ?

Amycus. That shalt thou learn when thirst has parched thy shrivelled lips.

Polydeuces. Will silver buy the boon, or with what price, prithee, may we gain thy leave ?

Amycus. Put up thy hands and stand in single combat, man to man.

Polydeuces. A boxing-match, or is kicking fair, when we meet eye to eye?

Amycus. Do thy best with thy fists and spare not thy skill !

Polydeuces. And who is the man on whom I am to lay my hands and gloves ?

Amycus. Thou see'st him close enough, the boxer will not prove a maiden !

Polydeuces. And is the prize ready, for which we two must fight ?

Amycus. Thy man shall I be called (shouldst thou win), or thou mine, if I be victor.

Polydeuces. On such terms fight the red-crested birds of the game.

Amycus. Well, be we like birds or lions, we shall fight for no other stake.

So Amycus spoke, and seized and blew his hollow shell, and speedily the long-haired Bebryces gathered beneath the shadowy planes,

I

at the blowing of the shell. And in likewise
did Castor, eminent in war, go forth and sum-
mon all the heroes from the Magnesian ship.
And the champions, when they had strength-
ened their fists with the stout ox-skin gloves,
and bound long leathern thongs about their
arms, stepped into the ring, breathing slaughter
against each other. Then had they much ado,
in that assault,—which should have the sun's
light at his back. But by thy skill, Polydeuces,
thou didst outwit the giant, and the sun's rays
fell full on the face of Amycus. Then came he
eagerly on in great wrath and heat, making
play with his fists, but the son of Tyndarus
smote him on the chin as he charged, madden-
ing him even more, and the giant confused the
fighting, laying on with all his weight, and
going in with his head down. The Bebryces
cheered their man, and on the other side the
heroes still encouraged stout Polydeuces, for
they feared lest the giant's weight, a match for
Tityus, might crush their champion in the
narrow lists. But the son of Zeus stood to
him, shifting his ground again and again, and
kept smiting him, right and left, and somewhat
checked the rush of the son of Posidon, for all
his monstrous strength. Then he stood reeling
like a drunken man under the blows, and spat
out the red blood, while all the heroes together
raised a cheer, as they marked the woful bruises
about his mouth and jaws, and how, as his face
swelled up, his eyes were half closed. Next,
the prince teased him, feinting on every side,

but seeing now that the giant was all abroad,
he planted his fist just above the middle of the
nose, beneath the eyebrows, and skinned all the
brow to the bone. Thus smitten, Amycus lay
stretched on his back, among the flowers and
grasses. There was fierce fighting when he
arose again, and they bruised each other well,
laying on with the hard weighted gloves ; but
the champion of the Bebryces was always play-
ing on the chest, and outside the neck, while
unconquered Polydeuces kept smashing his
foeman's face with ugly blows. The giant's
flesh was melting away in his sweat, till from a
huge mass he soon became small enough, but
the limbs of the other waxed always stronger,
and his colour better, as he warmed to his
work.

How then, at last, did the son of Zeus lay
low the glutton ? say goddess, for thou knowest,
but I, who am but the interpreter of others,
will speak all that thou wilt, and in such wise
as pleases thee.

Now behold the giant was keen to do some
great feat, so with his left hand he grasped the
left of Polydeuces, stooping slantwise from his
onset, while with his other hand he made his
effort, and drove a huge fist up from his right
haunch. Had his blow come home, he would
have harmed the King of Amyclae, but he
slipped his head out of the way, and then with
his strong hand struck Amycus on the left
temple, putting his shoulder into the blow.
Quick gushed the black blood from the gaping

temple, while Polydeuces smote the giant's mouth with his left, and the close-set teeth rattled. And still he punished his face with quick-repeated blows, till the cheeks were fairly pounded. Then Amycus lay stretched all on the ground, fainting, and held out **both his** hands, to show that he declined the fight, for he was near to death.

There then, despite thy victory, didst thou work him no insensate wrong, O boxer Polydeuces, but to thee he swore a mighty oath, calling his sire Posidon from the deep, that assuredly never again would he be violent to strangers.

Thee have I hymned, my prince ; but thee now, Castor, will I sing, O son of Tyndarus, O lord of the swift steeds, O **wielder** of the spear, thou that wearest the corselet of bronze.

Now these twain, the sons of Zeus, had seized and were bearing away the two daughters of Lycippus, and eagerly in sooth these two other brethren were pursuing them, the sons of Aphareus, even they that should soon have been the bridegrooms, — Lynceus and mighty Idas. But when they were come to the tomb of the dead Aphareus, then forth **from** their chariots they all sprang together, and set upon each other, under the weight of their spears and hollow shields. But Lynceus again spake, and shouted loud from under his vizor :—

'Sirs, wherefore desire ye battle, and how

are ye thus violent to win the brides of others
with naked swords in your hands. To us, behold, did Leucippus betroth these his daughters
long before; to us this bridal is by oath con-
firmed. And ye did not well, in that to win
the wives of others ye perverted him with gifts
of oxen, and mules, and other wealth, and so
won wedlock by bribes. Lo many a time, in
face of both of you, I have spoken thus, I that
am not a man of many words, saying,— "Not
thus, dear friends, does it become heroes to
woo their wives, wives that already have bridegrooms betrothed. Lo Sparta is wide, and
wide is Elis, a land of chariots and horses,
and Arcadia rich in sheep, and there are the
citadels of the Achaeans, and Messenia, and
Argos, and all the sea-coast of Sisyphus.
There be maidens by their parents nurtured,
maidens countless, that lack not aught in
wisdom or in comeliness. Of these ye may
easily win such as ye will, for many are willing
to be the fathers-in-law of noble youths, and ye
are the very choice of heroes all, as your
fathers were, and all your father's kin, and all
your blood from of old. But, friends, let this
our bridal find its due conclusion, and for you
let all of us seek out another marriage."

'Many such words I would speak, but the
wind's breath bare them away to the wet wave
of the sea, and no favour followed with my
words. For ye twain are hard and ruthless,—
nay, but even now do ye listen, for ye are our
cousins, and kin by the father's side. But if

your heart yet lusts for war, and with blood we
must break up the kindred strife, and end the
feud,[1] then Idas and his cousin, mighty Polydeuces, shall hold their hands and abstain from
battle, but let us twain, Castor and I, the
younger born, try the ordeal of war! Let us
not leave the heaviest of grief to our fathers!
Enough is one slain man from a house, but the
others will make festival for all their friends,
and will be bridegrooms, not slain men, and
will wed these maidens. Lo, it is fitting with
light loss to end a great dispute.'

So he spake, and these words the gods were
not to make vain. For the elder pair laid
down their harness from their shoulders on the
ground, but Lynceus stepped into the midst,
swaying his mighty spear beneath the outer
rim of his shield, and even so did Castor sway
his spear-points, and the plumes were nodding
above the crests of each. With the sharp
spears long they laboured and tilted at each
other, if perchance they might anywhere spy a
part of the flesh unarmed. But ere either was
wounded the spear-points were broken, fast
stuck in the linden shields. Then both drew
their swords from the sheaths, and again
devised each the other's slaying, and there was
no truce in the fight. Many a time did Castor
smite on broad shield and horse-hair crest, and
many a time the keen-sighted Lynceus smote
upon his shield, and his blade just shore the

---

[1] Or ἔγχεα λοῦσαι, 'wash the spears,' as in the Zulu
idiom.

scarlet plume. Then, as he aimed the sharp
sword at the left knee, Castor drew back with
his left foot, and hacked the fingers off the
hand of Lynceus. Then he being smitten cast
away his sword, and turned swiftly to flee to
the tomb of his father, where mighty Idas lay,
and watched this strife of kinsmen. But the
son of Tyndarus sped after him, and drove the
broad sword through bowels and navel, and
instantly the bronze cleft all in twain, and Lynceus bowed, and on his face he lay fallen on
the ground, and forthwith heavy sleep rushed
down upon his eyelids.

Nay, nor that other of her children did
Laocoosa see, by the hearth of his fathers, after
he had fulfilled a happy marriage. For lo,
Messenian Idas did swiftly break away the
standing stone from the tomb of his father
Aphareus, and now he would have smitten the
slayer of his brother, but Zeus defended him
and drave the polished stone from the hands of
Idas, and utterly consumed him with a flaming
thunderbolt.

Thus it is no light labour to war with the
sons of Tyndarus, for a mighty pair are they,
and mighty is he that begat them.

Farewell, ye children of Leda, and all goodly
renown send ye ever to our singing. Dear are
all minstrels to the sons of Tyndarus, and to
Helen, and to the other heroes that sacked
Troy in aid of Menelaus.

For you, O princes, the bard of Chios
wrought renown, when he sang the city of

Priam, and the ships of the Achaeans, and the
Ilian war, and Achilles, a tower of battle.
And to you, in my turn, the charms of the
clear-voiced Muses, even all that they can give,
and all that my house has in store, these do I
bring.   The fairest meed of the gods is song.

:

IDYL XXIII

THE VENGEANCE OF LOVE

*A lover hangs himself at the gate of his obdurate darling
who, in turn, is slain by a statue of Love.
This poem is not attributed with much certainty to
Theocritus, and is found in but a small proportion
of manuscripts.*

A LOVE-SICK youth pined for an unkind love,
beautiful in form, but fair no more in mood.
The beloved hated the lover, and had for him
no gentleness at all, and knew not Love, how
mighty a God is he, and what a bow his hands
do wield, and what bitter arrows he dealeth at
the young. Yea, in all things ever, in speech
and in all approaches, was the beloved unyield-
ing. Never was there any assuagement of
Love's fires, never was there a smile of the
lips, nor a bright glance of the eyes, never a
blushing cheek, nor a word, nor a kiss that
lightens the burden of desire. Nay, as a
beast of the wild wood hath the hunters in
watchful dread, even so did the beloved in all
things regard the man, with angered lips, and
eyes that had the dreadful glance of fate, and

the whole face was answerable to this wrath, the colour fled from it, sicklied o'er with wrathful pride. Yet even thus was the loved one beautiful, and the lover was the more moved by this haughtiness. At length he could no more endure so fierce a flame of the Cytherean, but drew near and wept by the hateful dwelling, and kissed the lintel of the door, and thus he lifted up his voice:

'O cruel child, and hateful, thou nursling of some fierce lioness, O child all of stone, unworthy of love; I have come with these my latest gifts to thee, even this halter of mine; for, child, I would no longer anger thee and work thee pain. Nay, I am going where thou hast condemned me to fare, where, as men say, is the path, and there the common remedy of lovers, the River of Forgetfulness. Nay, but were I to take and drain with my lips all the waters thereof, not even so shall I quench my yearning desire. And now I bid my farewell to these gates of thine.

'Behold I know the thing that is to be.

'Yea, the rose is beautiful, and Time he withers it; and fair is the violet in spring, and swiftly it waxes old; white is the lily, it fadeth when it falleth; and snow is white, and melteth after it hath been frozen. And the beauty of youth is fair, but lives only for a little season.

'That time will come when thou too shalt love, when thy heart shall burn, and thou shalt weep salt tears.

'But, child, do me even this last favour; when thou comest forth, and see'st me hanging in thy gateway,—pass me not careless by, thy hapless lover, but stand, and weep a little while; and when thou hast made this libation of thy tears, then loose me from the rope, and cast over me some garment from thine own limbs, and so cover me from sight; but first kiss me for that latest time of all, and grant the dead this grace of thy lips.

'Fear me not, I cannot live again, no, not though thou shouldst be reconciled to me, and kiss me. A tomb for me do thou hollow, to be the hiding-place of my love, and if thou departest, cry thrice above me,—

O friend, thou liest low!

And if thou wilt, add this also,—

Alas, my true friend is dead!

'And this legend do thou write, that I will scratch on thy walls,—

*This man Love slew! Wayfarer, pass not
   heedless by,
But stand, and say, "he had a cruel darling."'*

Therewith he seized a stone, and laid it against the wall, as high as the middle of the doorposts, a dreadful stone, and from the lintel he fastened the slender halter, and cast the noose about his neck, and kicked away the support from under his foot, and there was he hanged dead.

But the beloved opened the door, and saw
the dead man hanging there in the court, unmoved of heart, and tearless for the strange,
woful death ; but on the dead man were all
the garments of youth defiled.   Then forth
went the beloved to the contests of the
wrestlers, and there was heart-set on the delightful bathing-places, and even thereby en-
countered the very God dishonoured, for Love
stood on a pedestal of stone above the waters.[1]
And lo, the statue leaped, and slew that cruel
one, and the water was red with blood, but
the voice of the slain kept floating to the brim.

Rejoice, ye lovers, for he that hated is slain.
Love, all ye beloved, for the God knoweth how
to deal righteous judgment.

[1] In line 57 for τῆλε read Wordsworth's conjecture
τῇδε = ἐνταῦθα.

THE INFANT HERACLES

This poem describes the earliest feat of Heracles, the slaying of the snakes sent against him by Hera, and gives an account of the hero's training. The vivacity and tenderness of the pictures of domestic life, and the minute knowledge of expiatory ceremonies seem to stamp this idyl as the work of Theocritus. As the following poem also deals with an adventure of Heracles, it seems not impossible that Theocritus wrote, or contemplated writing, a Heraclean epic, in a series of idyls.

WHEN Heracles was but ten months old, the lady of Midea, even Alcmena, took him, on a time, and Iphicles his brother, younger by one night, and gave them both their bath, and their fill of milk, then laid them down in the buckler of bronze, that goodly piece whereof Amphitryon had strippen the fallen Pterelaus. And then the lady stroked her children's heads, and spoke, saying :—

'Sleep, my little ones, a light delicious sleep ; sleep, soul of mine, two brothers, babes unharmed ; blessed be your sleep, and blessed may ye come to the dawn.'

So speaking she rocked the huge shield, and
in a moment sleep laid hold on them.

But when the *Bear* at midnight wheels westward over against *Orion* that shows his mighty
shoulder, even then did crafty Hera send forth
two monstrous things, two snakes bristling up
their coils of azure ; against the broad threshold, where are the hollow pillars of the house-
door she urged them ; with intent that they
should devour the young child Heracles.
Then these twain crawled forth, writhing their
ravenous bellies along the ground, and still
from their eyes a baleful fire was shining as
they came, and they spat out their deadly
venom.   But when with their flickering tongues
they were drawing near the children, then
Alcmena's dear babes wakened, by the will of
Zeus that knows all things, and there was a
bright light in the chamber.   Then truly one
child, even Iphicles, screamed out straightway,
when he beheld the hideous monsters above
the hollow shield, and saw their pitiless fangs,
and he kicked off the woollen coverlet with his
feet, in his eagerness to flee.   But Heracles
set his force against them, and grasped them
with his hands, binding them both in a grievous bond, having got them by the throat,
wherein lies the evil venom of baleful snakes,
the venom detested even by the gods.   Then
the serpents, in their turn, wound with their
coils about the young child, the child unweaned,
that wept never in his nursling days ; but
again they relaxed their spines in stress of

pain, and strove to find some issue from the grasp of iron.

Now Alcmena heard the cry, and wakened first,—

'Arise, Amphitryon, for numbing fear lays hold of me: arise, nor stay to put shoon beneath thy feet! Hearest thou not how loud the younger child is wailing? Mark'st thou not that though it is the depth of **the** night, the walls are all plain to see as in the clear dawn?[1] There is some strange thing I trow within the house, there is, my dearest lord!'

Thus she spake, and at his wife's bidding he stepped down out of his bed, and made for his richly dight sword that he kept always hanging on its pin above his bed of cedar. Verily he was reaching out for his new-woven belt, lifting with the other hand the mighty sheath, a work of lotus wood, when lo, the wide chamber was filled again with night. Then he cried aloud on his thralls, who were drawing the deep breath of sleep,—

'Lights! Bring lights as quick as may be from the hearth, my thralls, and thrust back **the** strong bolts of the doors. Arise, ye serving-men, stout of heart, 'tis the master calls.'

Then quick the serving-men came speeding with torches burning, and the house waxed full

---

[1] Odyssey, xix. 36 seq. (Reading ἄπερ not ἄτερ.)
'Father, surely a great marvel is this that I behold with mine eyes; meseems, at least, that the walls of the hall . . . are bright as it were with flaming fire' . . .
'Lo! this is the wont of the gods that hold Olympus.'

as each man hasted along. Then truly when
they saw the young child Heracles clutching
the snakes twain in his tender grasp, they all
cried out and smote their hands together. But
he kept showing the creeping things to his
father, Amphitryon, and leaped on high in his
childish glee, and laughing, at his father's
feet he laid them down, the dread monsters
fallen on the sleep of death. Then Alcmena
in her own bosom took and laid Iphicles,
dry-eyed and wan with fear;[1] but Amphitryon, placing the other child beneath a lamb's-
wool coverlet, betook himself again to his bed,
and gat him to his rest.

The cocks were now but singing their third
welcome to the earliest dawn, when Alcmena
called forth Tiresias, the seer that cannot lie,
and told him of the new portent, and bade him
declare what things should come to pass.

'Nay, and even if the gods devise some
mischief, conceal it not from me in ruth and
pity ; and how that mortals may not escape
the doom that Fate speeds from her spindle,
O soothsayer Euerides, I am teaching thee,
that thyself knowest it right well.'

Thus spake the Queen, and thus he answered her :

'Be of good cheer, daughter of Perseus,
woman that hast borne the noblest of children
[and lay up in thy heart the better of the
things that are to be]. For by the sweet light
that long hath left mine eyes, I swear that

---

[1] ξηρὸν, *prae timore non lacrymantem* (Paley).

many Achaean women, as they card the soft
wool about their knees, shall sing at eventide,
of Alcmena's name, and thou shalt be honourable among the women of Argos. Such a man,
even this thy son, shall mount to the starry fir-
mament, the hero broad of breast, the master
of all wild beasts, and of all mankind. Twelve
labours is he fated to accomplish, and thereafter to dwell in the house of Zeus, but all his
mortal part a Trachinian pyre shall possess.

'And the son of the Immortals, by virtue of
his bride, shall he be called, even of them that
urged forth these snakes from their dens to
destroy the child. Verily that day shall come
when the ravening wolf, beholding the fawn in
his lair, will not seek to work him harm.

'But lady, see that thou hast fire at hand,
beneath the embers, and let make ready dry
fuel of gorse, or thorn, or bramble, or pear
boughs dried with the wind's buffeting, and on
the wild fire burn these serpents twain, at midnight, even at the hour when they would have
slain thy child. But at dawn let one of thy
maidens gather the dust of the fire, and bear
and cast it all, every grain, over the river from
the brow of the broken cliff,[1] beyond the march
of your land, and return again without looking

---

[1] Reading, after Fritzsche, ῥωγάδος ἐκ πέτρας. We
should have expected the accursed ashes (like those of
Wyclif) to be thrown *into* the river ; cf. Virgil, Ecl.
viii. 101, ' Fer cineres, Amarylli, foras, rivoque fluenti
transque caput jace nec respexeris.' Virgil's knowledge
of these observances was not inferior to that of Theocritus.

behind. Then cleanse your house with the
fire of unmixed sulphur first, and then, as is
ordained, with a filleted bough sprinkle holy
water over all, mingled with salt.[1] And to
Zeus supreme, moreover, do ye sacrifice a
young boar, that ye may ever have the mastery
over all your enemies.'

So spake he, and thrust back his ivory chair,
and departed, even Tiresias, despite the weight
of all his many years.

But Heracles was reared under his mother's
care, like some young sapling in a garden close,
being called the son of Amphitryon of Argos.
And the lad was taught his letters by the ancient
Linus, Apollo's son, a tutor ever watchful. And
to draw the bow, and send the arrow to the
mark did Eurytus teach him, Eurytus rich in
wide ancestral lands. And Eumolpus, son of
Philammon, made the lad a minstrel, and formed
his hands to the boxwood lyre. And all the
tricks wherewith the nimble Argive cross-but-
tockers give each other the fall, and all the
wiles of boxers skilled with the gloves, and all
the art that the rough and tumble fighters have
sought out to aid their science, all these did
Heracles learn from Harpalacus of Phanes, the
son of Hermes. Him no man that beheld,
even from afar, would have confidently met as
a wrestler in the lists, so grim a brow overhung
his dreadful face. And to drive forth his horses
'neath the chariot, and safely to guide them

---

[1] Reading ἐστεμμένῳ If ἐστεμμένον is read, the
phrase will mean ' pure brimming water.'

round the goals, with the naves of the wheels
unharmed, Amphitryon taught his son in his
loving-kindness, Amphitryon himself, for many
a prize had he borne away from the fleet races
in Argos, pasture-land of steeds, and unbroken
were the chariots that he mounted, till time
loosened their leathern thongs.

But to charge with spear in rest, against a
foe, guarding, meanwhile, his back with the
shield, to bide the biting swords, to order a
company, and to measure, in his onslaught, the
ambush of foemen, and to give horsemen the
word of command, he was taught by knightly
Castor. An outlaw came Castor out of Argos,
when Tydeus was holding all the land and all
the wide vineyards, having received Argos, a
land of steeds, from the hand of Adrastus. No
peer in war among the demigods had Castor,
till age wore down his youth.

Thus did his dear mother let train Heracles,
and the child's bed was made hard by his
father's ; a lion's skin was the coverlet he loved ;
his dinner was roast meat, and a great Dorian
loaf in a basket, a meal to satisfy a delving
hind. At the close of day he would take a
meagre supper that needed no fire to the cook-
ing, and his plain kirtle fell no lower than the
middle of his shin.

IDYL XXV

HERACLES THE LION-SLAYER

This is another idyl of the epic sort. The poet's interest in the details of the rural life, and in the description of the herds of King Augeas, seem to mark it as the work of Theocritus. It has, however, been attributed by learned conjecture to various writers of an older age. The idyl, or fragment, is incomplete. Heracles visits the herds of Augeas (to clean their stalls was one of his labours), and, after an encounter with a bull, describes to the king's son his battle with the lion of Nemea.

. . . Him answered the old man, a husband-man that had the care of the tillage, ceasing a moment from the work that lay betwixt his hands—

'Right readily will I tell thee, stranger, concerning the things whereof thou inquirest, for I revere the awful wrath of Hermes of the road-side. Yea he, they say, is of all the heavenly Gods the most in anger, if any deny the way-farer that asks eagerly for the way.

'The fleecy flocks of the king Augeas feed not all on one pasture, nor in one place, but some there be that graze by the river-banks

round Elisus, and some by the sacred stream
of divine Alpheius, and some by Buprasium
rich in clusters of the vine, and some even in
this place. And behold, the pens for each herd
after its kind are builded apart. Nay, but for
all the herds of Augeas, overflowing as they be,
these pasture lands are ever fresh and flowering,
around the great marsh of Peneus, for with
herbage honey-sweet the dewy water-meadows
are ever blossoming abundantly, and this fodder
it is that feeds the strength of horned kine.
And this their steading, on thy right hand
stands all plain to view, beyond the running
river, there, where the plane-trees grow luxuriant, and the green wild olive, a sacred grove,
O stranger, of Apollo of the pastures, a God
most gracious unto prayer. Next thereto are
builded long rows of huts for the country folk,
even for us that do zealously guard the great
and marvellous wealth of the king ; casting in
season the seed in fallow lands, thrice, ay, and
four times broken by the plough. As for the
marches, truly, the ditchers know them, men of
many toils, who throng to the wine-press at the
coming of high summer tide. For, behold, all
this plain is held by gracious Augeas, and the
wheat-bearing plough-land, and the orchards
with their trees, as far as the upland farm of
the ridge, whence the fountains spring ; over
all which lands we go labouring, the whole day
long, as is the wont of thralls that live their
lives among the fields.

'But, prithee, tell thou me, in thy turn (and

for thine own gain it will be), whom comest
thou hither to seek; in quest, perchance, of
Augeas, or one of his servants? Of all these
things, behold, I have knowledge, and could
tell thee plainly, for methinks that thou, for thy
part, comest of no churlish stock, nay, nor hath
thy shape aught of the churl, so excellent in
might shows thy form. Lo, now, even such are
the children of the immortal Gods among mortal
men.' Then the mighty son of Zeus answered
him, saying—

'Yea, old man, I fain would see Augeas,
prince of the Epeans, for truly 'twas need of
him that brought me hither. If he abides at
the town with his citizens, caring for his people,
and settling the pleas, do thou, old man, bid
one of the servants to guide me on the way,
a head-man of the more honourable sort in
these fields, to whom I may both tell my desire,
and learn in turn what I would, for God has
made all men dependent, each on each.'

Then the old man, the worthy husbandman,
answered him again—

'By the guidance of some one of the immortals hast thou come hither, stranger, for
verily all that thou requirest hath quickly been
fulfilled. For hither hath come Augeas, the
dear son of Helios, with his own son, the strong
and princely Phyleus. But yesterday he came
hither from the city, to be overseeing after
many days his substance, that he hath un-
counted in the fields. Thus do even kings in
their inmost hearts believe that the eye of the

master makes the house more prosperous. Nay come, let us hasten to him, and I will lead thee to our dwelling, where methinks we shall find the king.'

So he spake, and began to lead the way, but in his mind, as he marked the lion's hide, and the club that filled the stranger's fist, the old man was deeply pondering as to whence he **came**, and ever he was eager to inquire of him. But back again he kept catching the word as it rose to his lips, in fear lest he should speak somewhat out of season (his companion being in haste) for hard it is to know another's mood.

Now as they began to draw nigh, the dogs from afar were instantly aware of them, both by the scent, and by the sound of footsteps, and, yelling furiously, they charged from **all** sides against Heracles, son of Amphitryon, while with faint yelping, on the other side, they greeted the old man, and fawned around him. But he just lifted stones from the ground,[1] and scared them away, and, raising his voice, he right roughly chid them all, and made them cease from their yelping, being glad in his heart withal for that they guarded his dwelling, even when he was afar. Then thus he spake—

'Lo, what a comrade for men have the Gods, the lords of all, made in this creature, how mindful is he! If he had but so much wit within him as to know against whom he should

---

[1] Reading ὅσσον.

rage, and with whom he should forbear, no beast in the world could vie with his deserts. But now he is something over-fierce and blindly furious.'

So he spake, and they hastened, and came even to that dwelling whither they were faring.

Now Helios had turned his steeds to the west, bringing the late day, and the fatted sheep came up from the pastures to the pens and folds. Next thereafter the kine approaching, ten thousand upon ten thousand, showed for multitude even like the watery clouds that roll forward in heaven under the stress of the South Wind, or the Thracian North (and countless are they, and ceaseless in their airy passage, for the wind's might rolls up the rear as numerous as the van, and hosts upon hosts again are moving in infinite array), even so many did herds upon herds of kine move ever forwards. And, lo, the whole plain was filled, and all the ways, as the cattle fared onwards, and the rich fields could not contain their lowing, and the stalls were lightly filled with kine of trailing feet, and the sheep were being penned in the folds.

There no man, for lack of labour, stood idle by the cattle, though countless men were there, but one was fastening guards of wood, with shapely thongs, about the feet of the kine, that he might draw near and stand by, and milk them. And another beneath their mothers kind was placing the calves right eager to drink of the sweet milk. Yet another held a

milking pail, while his fellow was fixing the
rich cheese, and another led in the bulls apart
from the cows. Meanwhile Augeas was going
round all the stalls, and marking the care his
herdsmen bestowed upon all that was his.
And the king's son, and the mighty, deep-
pondering Heracles, went along with the **king,**
as he passed through his great possessions.
Then though he bore a stout spirit in his heart,
and a mind stablished always imperturbable,
yet the son of Amphitryon still marvelled out
of measure, as he beheld these countless troops
of cattle. Yea none would have deemed or
believed that the substance of one man could
be so vast, nay, nor ten men's wealth, were
they the richest in sheep of all the kings in the
world. But Helios to his son gave this **gift**
pre-eminent, namely to abound in flocks far
above all other men, and Helios himself did
ever and always give increase to the cattle, for
upon his herds came no disease, of them that
always minish the herdman's toil. But always
more in number waxed the horned kine, and
goodlier, year by year, for verily they all
brought forth exceeding abundantly, and never
cast their young, and chiefly bare heifers.

With the kine went continually three hundred
bulls, white-shanked, and curved of horn,—and
two hundred others, red cattle,—and all these
already were of an age to mate with the kine.
Other twelve bulls, again, besides these, went
together in a herd, being sacred to Helios.
They were white as swans, and shone among

all the herds of trailing gait. And these disdaining the herds grazed still on the rich herbage in the pastures, and they were exceeding high of heart. And whensoever the swift wild beasts came down from the rough oakwood to the plain, to seek the wilder cattle, afield went these bulls first to the fight, at the smell of the savour of the beasts, bellowing fearfully, and glancing slaughter from their brows.

Among these bulls was one pre-eminent for strength and might, and for reckless pride, even the mighty Phaethon, that all the herdsmen still likened to a star, because he always shone so bright when he went among the other cattle, and was right easy to be discerned. Now when this bull beheld the dried skin of the fierce-faced lion, he rushed against the keen-eyed Heracles himself, to dash his head and stalwart front against the sides of the hero. Even as he charged, the prince forthwith grasped him with strong hand by the left horn, and bowed his neck down to the ground, puissant as he was, and, with the weight of his shoulder, crushed him backwards, while clear stood out the strained muscle over the sinews on the hero's upper arm. Then marvelled the king himself, and his son, the warlike Phyleus, and the herdsmen that were set over the horned kine,—when they beheld the exceeding strength of the son of Amphitryon.

Now these twain, even Phyleus and mighty Heracles, left the fat fields there, and were making for the city. But just where they

entered on the highway, after quickly speeding over the narrow path that stretched through
the vineyard from the farmhouses, a dim path
through the green wood, thereby the dear son
of Augeas bespake the child of supreme Zeus,
who was behind him, slightly turning his head
over his right shoulder,

'Stranger, long time ago I heard a tale,
which, as of late I guess, surely concerneth
thee. For there came hither, in his wayfaring
out of Argos, a certain young Achaean, from
Helicé, by the seashore, who verily told a tale
and that among many Epeians here,—how,
even in his presence, a certain Argive slew a
wild beast, a lion dread, a curse of evil omen
to the country folk. The monster had its
hollow lair by the grove of Nemean Zeus, but
as for him that slew it, I know not surely
whether he was a man of sacred Argos, there,
or a dweller in Tiryns city, or in Mycenae, as
he that told the tale declared. By birth, how-
beit, he said (if rightly, I recall it) that the hero
was descended from Perseus. Methinks that
none of the Aegialeis had the hardihood for
this deed save thyself; nay, the hide of the
beast that covers thy sides doth clearly pro-
claim the mighty deed of thy hands. But
come now, hero, tell thou me first, that truly
I may know, whether my foreboding be right
or wrong,—if thou art that man of whom the
Achaean from Helicé spake in our hearing,
and if I read thee aright. Tell me how single-
handed thou didst slay this ruinous pest, and

how it came to the well-watered ground of Nemea,
for not in Apis couldst thou find,—not though
thou soughtest after it,—so great a monster.
For the country feeds no such large game, but
bears, and boars, and the pestilent race of
wolves. Wherefore all were in amaze that
listened to the story, and there were some who
said that the traveller was lying, and pleasing
them that stood by with the words of an idle
tongue.'

Thus Phyleus spake, and stepped out of the
middle of the road, that there might be space
for both to walk abreast, and that so he might
hear the more easily the words of Heracles who
now came abreast with him, and spake thus,

'O son of Augeas, concerning that whereof
thou first didst ask me, thyself most easily hast
discerned it aright. Nay then, about this monster I will tell thee all, even how all was done,
—since thou art eager to hear,—save, indeed,
as to whence he came, for, many as the Argives
be, not one can tell that clearly. Only we guess
that some one of the Immortals, in wrath for
sacrifice unoffered, sent this bane against the
children of Phoroneus. For over all the men
of Pisa the lion swept, like a flood, and still
ravaged insatiate, and chiefly spoiled the Bem-
binaeans, that were his neighbours, and endured
things intolerable.

'Now this labour did Eurystheus enjoin on
me to fulfil the first of all, and bade me slay
the dreadful monster. So I took my supple
bow, and hollow quiver full of arrows, and set

forth; and in my other hand I held my stout
club, well balanced, and wrought, with unstripped
bark, from a shady wild olive-tree, that I myself
had found, under sacred Helicon, and dragged
up the whole tree, with the bushy roots. **But**
when I **came to the** place whereby the lion
abode, even then I grasped my bow and slipped
the string up to the curved tip, and straightway
laid thereon the bitter arrow. Then I cast my
eyes on every side, spying for the baneful mon-
ster, if perchance I might see him, or ever he
saw me. It was now midday, and nowhere
might I discern the tracks of the monster, nor
hear his roaring. Nay, nor was there one
man to be seen with the cattle, and the tillage
through all the furrowed lea, of whom I might
inquire, but wan fear still held them all within
the homesteads. Yet I stayed not in my going,
as I quested through the deep-wooded hill, till
I beheld him, and instantly essayed my prowess.
Now early in the evening he was making for
his lair, full fed with blood and flesh, and all
his bristling mane was dashed with carnage,
and his fierce face, and his breast, and still
with his tongue he kept licking his bearded
chin. Then instantly I hid me in the dark
undergrowth, on the wooded hill, awaiting his
approach, and as he came nearer I smote him
on the left flank, but all in vain, for naught did
the sharp arrow pierce through his flesh, but
leaped back, and fell on the green grass. Then
quickly he raised his tawny head from the
ground, in amaze, glancing all around with

his eyes, and with jaws distent he showed his ravenous teeth.   Then I launched against him another shaft from the string, in wrath that the former flew vainly from my hand, and I smote him right in the middle of the breast, where the lung is seated, yet not even so did the cruel arrow sink into his hide, but fell before his feet, in vain, to no avail.   Then for the third time was I making ready to draw my bow again, in great shame and wrath, but the furious beast glanced his eyes around, and spied me.   With his long tail he lashed his flanks, and straightway bethought him of battle. His neck was clothed with wrath, and his tawny hair bristled round his lowering brow, and his spine was curved like a bow, his whole force being gathered up from under towards his flanks and loins.   And as when a wainwright, one skilled in many an art, doth bend the saplings of seasoned fig-tree, having first tempered them in the fire, to make tires for the axles of his chariot, and even then the fig-tree wood is like to leap from his hands in the bending, and springs far away at a single bound, even so the dread lion leaped on me from afar, huddled in a heap, and keen to glut him with my flesh.   Then with one hand I thrust in front of me my arrows, and the double folded cloak from my shoulder, and with the other raised the seasoned club above my head, and drove at his crest, and even on the shaggy scalp of the insatiate beast brake my grievous cudgel of wild olive-tree.   Then or ever he reached me, he fell from his flight, on to the ground, and stood on trembling feet, with wagging head, for darkness gathered about both his eyes, his brain being shaken in his skull with the violence of the blow. Then when I marked how he was distraught with the grievous torment, or ever he could turn and gain breath again, I fell on him, and seized him by the column of his stubborn neck. To earth I cast my bow, and woven quiver, and strangled him with all my force, gripping him with stubborn clasp from the rear, lest he should rend my flesh with his claws, and I sprang on him and kept firmly treading his hind feet into the soil with my heels, while I used his sides to guard my thighs, till I had strained his shoulders utterly, then lifted him up, all breathless,—and Hell took his monstrous life.

'And then at last I took thought how I should strip the rough hide from the dead beast's limbs, a right hard labour, for it might not be cut with steel, when I tried, nor stone, nor with aught else.[1] Thereon one of the Immortals put into my mind the thought to cleave the lion's hide with his own claws. With these I speedily flayed it off, and cast it about my limbs, for my defence against the brunt of wounding war.

'Friend, lo even thus befel the slaying of the Nemean Lion, that aforetime had brought many a bane on flocks and men.'

[1] Reading ἄλλῃ, as in Wordsworth's conjecture, instead of ὕλῃ.

This idyl narrates the murder of Pentheus, who was torn to pieces (after the Dionysiac Ritual) by his mother, Agave, and other Theban women, for having watched the celebration of the mysteries of Dionysus. It is still dangerous for an Australian native to approach the women of the tribe while they are celebrating their savage rites. The conservatism of Greek religion is well illustrated by Theocritus's apology for the truly savage revenge commemorated in the old Theban legend.

INO, and Autonoe, and Agave of the apple cheeks, — three bands of Maenads to the mountain - side they led, these ladies three. They stripped the wild leaves of a rugged oak, and fresh ivy, and asphodel of the upper earth, and in an open meadow they built twelve altars; for Semele three, and nine for Dionysus. The mystic cakes [1] from the mystic chest they had taken in their hands, and in silence had laid them on the altars of new-stripped boughs; so Dionysus ever taught the rite, and herewith was he wont to be well pleased.

Now Pentheus from a lofty cliff was watch-

---

[1] Reading ποπανεύματα.

ing all, deep hidden in an ancient lentisk bush, a plant of that land. Autonoe first beheld him, and shrieked a dreadful yell, and, rushing suddenly, with her feet dashed all confused the mystic things of Bacchus the wild. For these are things unbeholden of men profane. Frenzied was she, and then forthwith the others too were frenzied. Then Pentheus fled in fear, and they pursued after him, with raiment kirtled through the belt above the knee.

This much said Pentheus, 'Women, what would ye?' and thus answered Autonoe, 'That shalt thou straightway know, ere thou hast heard it.'

The mother seized her child's head, and cried loud, as is the cry of a lioness over her cubs, while Ino, for her part, set her heel on the body, and brake asunder the broad shoulder, shoulder-blade and all, and in the same strain wrought Autonoe. The other women tore the remnants piecemeal, and to Thebes they came, all bedabbled with blood, from the mountains bearing not Pentheus but repentance.[1]

I care for none of these things, nay, nor let another take thought to make himself the foe of Dionysus, not though one should suffer yet greater torments than these,—being but a child of nine years old or entering, perchance, on his tenth year. For me, may I be pure and holy, and find favour in the eyes of the pure!

From aegis-bearing Zeus hath this augury

---

[1] Πένθημα καὶ οὐ πενθῆα, a play on words difficult to retain in English. Compare Idyl xiii. line 74.

all honour, 'to the children of the godly the better fortune, but evil befall the offspring of the ungodly.'

'Hail to Dionysus, whom Zeus supreme brought forth in snowy Dracanus, when he had unburdened his mighty thigh, and hail to beautiful Semele: and to her sisters,—Cadmeian ladies honoured of all daughters of heroes,—who did this deed at the behest of Dionysus, a deed not to be blamed; let no man blame the actions of the gods.'

IDYL XXVII

THE WOOING OF DAPHNIS

*The authenticity of this idyl has been denied, partly
because the Daphnis of the poem is not identical in
character with the Daphnis of the first idyl. But
the piece is certainly worthy of a place beside the
work of Theocritus. The dialogue is here arranged
as in the text of Fritzsche.*

The Maiden. Helen the wise did Paris, another
neatherd, ravish!

Daphnis. 'Tis rather this Helen that kisses
her shepherd, even me![1]

The Maiden. Boast not, little satyr, for
kisses they call an empty favour.

Daphnis. Nay, even in empty kisses there is
a sweet delight.

The Maiden. I wash my lips, I blow away
from me thy kisses!

Daphnis. Dost thou wash thy lips? Then
give me them again to kiss!

The Maiden. 'Tis for thee to caress thy kine,
not a maiden unwed.

---

[1] The conjecture ἐμὰ δ' gives a good sense, *mea vero
Helena me potius ultra petit.*

Daphnis. Boast not, for swiftly thy youth flits by thee, like a dream.

The Maiden. The grapes turn to raisins, not wholly will the dry rose perish.

Daphnis. Come hither, beneath the wild olives, that I may tell thee a tale.

The Maiden. I will not come ; ay, ere now with a sweet tale didst thou beguile me.

Daphnis. Come hither, beneath the elms, to listen to my pipe !

The Maiden. Nay, please thyself, no woful tune delights me.

Daphnis. Ah maiden, see that thou too shun the anger of the Paphian.

The Maiden. Good-bye to the Paphian, let Artemis only be friendly !

Daphnis. Say not so, lest she smite thee, and thou fall into a trap whence there is no escape.

The Maiden. Let her smite an she will ; Artemis again would be my defender. Lay no hand on me ; nay, if thou do more, and touch me with thy lips, I will bite thee.[1]

Daphnis. From Love thou dost not flee, whom never yet maiden fled.

The Maiden. Escape him, by Pan, I do, but thou dost ever bear his yoke.

Daphnis. This is ever my fear lest he even give thee to a meaner man.

The Maiden. Many have been my wooers, but none has won my heart.

[1] Reading, as in Wordsworth's conjecture, μὴ 'πιβάλῃς τὰν χεῖρα, καὶ εἴ γ' ἔτι χεῖλος, ἀμύξω.

Daphnis. Yea I, out of many chosen, come **here** thy wooer.

The Maiden. Dear love, what can I do? Marriage has much annoy.

Daphnis. Nor pain nor sorrow has **marriage,** but mirth and dancing.

The Maiden. Ay, but **they say that women dread their lords.**

Daphnis. **Nay, rather** they always rule them, **—whom do** women fear?

The Maiden. Travail I dread, and sharp **is the shaft of** Eilithyia.

Daphnis. But thy queen is Artemis, that lightens labour.

The Maiden. But I fear childbirth, lest, perchance, I lose my beauty.

Daphnis. Nay, if thou bearest dear children thou wilt see the light revive in thy sons.

The **Maiden.** And **what** wedding gift dost **thou** bring me if I consent?

Daphnis. My whole flock, all my groves, and **all** my pasture land shall be thine.

The Maiden. Swear that thou wilt not win **me,** and then depart and leave me forlorn.

Daphnis. So help me Pan I would not leave **thee,** didst thou **even** choose to banish me!

The Maiden. Dost thou build me bowers, and a house, **and** folds **for flocks?**

Daphnis. **Yea,** bowers I build thee, **the** flocks I tend are fair.

The Maiden. But to my grey old father, **what** tale, ah what, shall I tell?

Daphnis. He will approve thy wedlock when he has heard my name.

The Maiden. Prithee, tell me that name of thine ; in a name there is often delight.

Daphnis. Daphnis am I, Lycidas is my father, and Nomaea is my mother.

The Maiden. Thou comest of men well-born, but there I am thy match.

Daphnis. I know it, thou art of high degree, for thy father is Menalcas.[1]

The Maiden. Show me thy grove, wherein is thy cattle-stall.

Daphnis. See here, how they bloom, my slender cypress-trees.

The Maiden. Graze on, my goats, I go to learn the herdsman's labours.

Daphnis. Feed fair, my bulls, while I show my woodlands to my lady !

The Maiden. What dost thou, little satyr ; why dost thou touch my breast ?

Daphnis. I will show thee that these earliest apples are ripe.[2]

The Maiden. By Pan, I swoon ; away, take back thy hand.

Daphnis. Courage, dear girl, why fearest thou me, thou art over fearful !

The Maiden. Thou makest me lie down by the water-course, defiling my fair raiment !

Daphnis. Nay, see, 'neath thy raiment fair I am throwing this soft fleece.

---

[1] Reading οἶδ', ἀκρατιμλη ἐσσι, with Fritzsche. Compare the conjecture of Wordsworth, 'Οὐδ' ἄκρα τί μὴ ἔσσι ;   [2] See Wordsworth's explanation.

The Maiden. Ah, ah, thou hast snatched my girdle too ; why hast thou loosed my girdle ?

Daphnis. These first-fruits I offer, a gift to the Paphian.

The Maiden. Stay, wretch, hark ; surely a stranger cometh ; nay, I hear a sound.

Daphnis. The cypresses do but whisper to each other of thy wedding.

The Maiden. Thou hast torn my mantle, and unclad am I.

Daphnis. Another mantle I will give thee, and an ampler far than thine.

The Maiden. Thou dost promise all things, but soon thou wilt not give me even a grain of salt.

Daphnis. Ah, would that I could give thee my very life.

The Maiden. Artemis, be not wrathful, thy votary breaks her vow.

Daphnis. I will slay a calf for Love, and for Aphrodite herself a heifer.

The Maiden. A maiden I came hither, a woman shall I go homeward.

Daphnis. Nay, a wife and a mother of children shalt thou be, no more a maiden.

So, each to each, in the joy of their young fresh limbs they were murmuring : it was the hour of secret love. Then she arose, and stole to herd her sheep ; with shamefast eyes she went, but her heart was comforted within her. And he went to his herds of kine, rejoicing in his wedlock.

IDYL XXVIII

This little piece of Aeolic verse accompanied the pre-
sent of a distaff, which Theocritus brought from
Syracuse to Theugenis, the wife of his friend Nicias,
the physician of Miletus. On the margin of a
translation by Longepierre (the famous book-collector),
Louis XIV wrote that this idyl is a model of
honourable gallantry.

O DISTAFF, thou friend of them that spin, gift
of grey-eyed Athene to dames whose hearts
are set on housewifery ; come, boldly come
with me to the bright city of Neleus, where the
shrine of the Cyprian is green 'neath its roof of
delicate rushes. Thither I pray that we may
win fair voyage and favourable breeze from
Zeus, that so I may gladden mine eyes with
the sight of Nicias my friend, and be greeted
of him in turn ;—a sacred scion is he of the
sweet-voiced Graces. And thee, distaff, thou
child of fair carven ivory, I will give into the
hands of the wife of Nicias : with her shalt
thou fashion many a thing, garments for men,
and much rippling raiment that women wear.
For the mothers of lambs in the meadows
might twice be shorn of their wool in the year,

with her goodwill, the dainty-ankled Theugenis,
so notable is she, and cares for all things that
wise matrons love.

Nay, not to houses slatternly or idle would I
have given thee, distaff, seeing that thou art a
countryman of mine. For that is thy native
city which Archias out of Ephyre founded, long
ago, the very marrow of the isle of the three
capes, a town of honourable men.[1] But now
shalt thou abide in the house of a wise physician,
who has learned all the spells that ward off
sore maladies from men, and thou shalt dwell
in glad Miletus with the Ionian people, to this
end,—that of all the townsfolk Theugenis may
have the goodliest distaff, and that thou mayst
keep her ever mindful of her friend, the lover
of song.

This proverb will each man utter that looks
on thee, 'Surely great grace goes with a little
gift, and all the offerings of friends are precious.'

[1] Syracuse.

IDYL XXIX

*This poem, like the preceding one, is written in the
    Aeolic dialect. The first line is quoted from
    Alcaeus. The idyl is attributed to Theocritus on
    the evidence of the scholiast on the* Symposium *of
    Plato.*

'WINE and truth,' dear child, says the proverb,
and in wine are we, and the truth we must tell.
Yes, I will say to thee all that lies in my soul's
inmost chamber. Thou dost not care to love
me with thy whole heart! I know, for I live
half my life in the sight of thy beauty, but all
the rest is ruined. When thou art kind, my
day is like the days of the Blessed, but when
thou art unkind, 'tis deep in darkness. How
can it be right thus to torment thy friend?
Nay, if thou wilt listen at all, child, to me, that
am thine elder, happier thereby wilt thou be, and
some day thou wilt thank me. Build one nest in
one tree, where no fierce snake can come ; for
now thou dost perch on one branch to-day, and
on another to-morrow, always seeking what is
new. And if a stranger see and praise thy
pretty face, instantly to him thou art more than
a friend of three years' standing, while him that

loved thee first thou holdest no higher than a
friend of three days. Thou savourest, methinks,
of the love of some great one ; nay, choose
rather all thy life ever to keep the love of one
that is thy peer. If this thou dost thou wilt be
well spoken of by thy townsmen, and Love will
never be hard to thee, Love that lightly van-
quishes the minds of men, and has wrought to
tenderness my heart that was of steel. Nay,
by thy delicate mouth I approach and beseech
thee, remember that thou wert younger yester-
year, and that we wax grey and wrinkled, or
ever we can avert it ; and none may recapture
his youth again, for the shoulders of youth are
winged, and we are all too slow to catch such
flying pinions.

Mindful of this thou shouldst be gentler,
and love me without guile as I love thee,
so that, when thou hast a manly beard,
we may be such friends as were Achilles and
Patroclus !

But, if thou dost cast all I say to the winds
to waft afar, and cry, in anger, ' Why, why, dost
thou torment me ?' then I,—that now for thy
sake would go to fetch the golden apples, or to
bring thee Cerberus, the watcher of the dead,
—would not go forth, didst thou stand at the
court-doors and call me. I should have rest
from my cruel love.

Fragment of the Berenice.

Athenaeus (vii. 284 A) quotes this fragment, which probably was part of a panegyric on Berenice, the mother of Ptolemy Philadelphus.

AND if any man that hath his livelihood from the salt sea, and whose nets serve him for ploughs, prays for wealth, and luck in fishing, let him sacrifice, at midnight, to this goddess, the sacred fish that they call 'silver white,' for that it is brightest of sheen of all,—then let the fisher set his nets, and he shall draw them full from the sea.

IDYL XXX

THE DEAD ADONIS

*This idyl is usually printed with the poems of Theo-
critus, but almost certainly is by another hand. I
have therefore ventured to imitate the metre of the
original.*

WHEN Cypris saw Adonis,
In death already lying
With all his locks dishevelled,
And cheeks turned wan and ghastly,
She bade the Loves attendant
To bring the boar before her.

And lo, **the** winged ones, fleetly
They scoured through all the wild wood;
The wretched boar they tracked him,
And bound and doubly bound him.
One fixed on him a halter,
And dragged him on, a captive,
Another drave him onward,
And smote him with his arrows.
But terror-struck the beast came,
For much he feared Cythere.

To him spake Aphrodite,—
' Of wild beasts all the vilest,
This thigh, by thee was 't wounded ?
Was 't thou that smote my lover ? '
To her the beast made answer—
' I swear to thee, Cythere,
By thee, and by thy lover,
Yea, and by these my fetters,
And them that do pursue me,—
Thy lord, thy lovely lover
I never willed to wound him ;
I saw him, like a statue,
And could not bide the burning,
Nay, for his thigh was naked,
And mad was I to kiss it,
And thus my tusk it harmed him.
Take these my tusks, O Cypris,
And break them, and chastise them,
For wherefore should I wear them,
These passionate defences ?
If this doth not suffice thee,
Then cut my lips out also,
Why dared they try to kiss him ? '

Then Cypris had compassion ;
She bade the Loves attendant
To loose the bonds that bound him.
From that day her he follows,
And flees not to the wild wood
But joins the Loves, and always
He bears Love's flame unflinching.

EPIGRAMS

The Epigrams of Theocritus are, for the most part, either inscriptions for tombs or cenotaphs, or for the pedestals of statues, or (as the third epigram) are short occasional pieces. Several of them are but doubtfully ascribed to the poet of the Idyls. The Greek has little but brevity in common with the modern epigram.

I

For a rustic Altar.

THESE dew-drenched roses and that tufted thyme are offered to the ladies of Helicon. And the dark-leaved laurels are thine, O Pythian Paean, since the rock of Delphi bare this leafage to thine honour. The altar this white-horned goat shall stain with blood, this goat that browses on the tips of the terebinth boughs.

II

For a Herdsman's Offering.

Daphnis, the white-limbed Daphnis, that pipes on his fair flute the pastoral strains offered to

Pan these gifts,—his pierced reed-pipes, his crook, a javelin keen, a fawn-skin, and the scrip wherein he was wont, on a time, to carry the apples of Love.

III

For a Picture.

Thou sleepest on the leaf-strewn ground, O Daphnis, resting thy weary limbs, and the stakes of thy nets are newly fastened on the hills. But Pan is on thy track, and Priapus, with the golden ivy wreath twined round his winsome head,—both are leaping at one bound into thy cavern. Nay, flee them, flee, shake off thy slumber, shake off the heavy sleep that is falling upon thee.

IV

Priapus.

When thou hast turned yonder lane, goat-herd, where the oak-trees are, thou wilt find an image of fig-tree wood, newly carven; three-legged it is, the bark still covers it, and it is earless withal, yet meet for the arts of Cypris. A right holy precinct runs round it, and a ceaseless stream that falleth from the rocks on every side is green with laurels, and myrtles, and fragrant cypress. And all around the place that child of the grape, the vine, doth flourish with its tendrils, and the merles in spring with their sweet songs utter their wood-notes wild, and the brown nightingales reply with their complaints, pouring from their bills the honey-sweet song. There, prithee, sit down and pray to **gracious Priapus, that I may** be delivered from **my love of Daphnis, and say** that **instantly thereon I will** sacrifice a fair kid. But **if** he refuse, ah then, should **I win Daphnis's** love, **I would** fain sacrifice three **victims,—and** offer **a** calf, a shaggy he-goat, and a lamb that I **keep** in the stall, and oh that graciously the god may hear my prayer.

V

The rural Concert.

Ah, in the Muses' **name, wilt thou play me** some **sweet air on** the **double flute, and I will** take up the harp, and touch **a note, and the** **neatherd** Daphnis will charm us the while, breathing music into his wax-bound pipe. And beside this rugged oak behind the cave will we stand, and rob the goat-foot Pan of his repose.

VI

The Dead are beyond hope.

Ah hapless Thyrsis, where is **thy gain,** shouldst thou lament till thy two eyes are consumed with tears? She has passed away,— the kid, the **youngling** beautiful, — she has

M

passed away to Hades. Yea, the jaws of the
fierce wolf have closed on her, and now the
hounds are baying, but what avail they when
nor bone nor cinder is left of her that is departed?

VII

For a statue of Asclepius.

Even to Miletus he hath come, the son of
Paeon, to dwell with one that is a healer of all
sickness, with Nicias, who even approaches
him day by day with sacrifices, and hath let
carve this statue out of fragrant cedar-wood;
and to Eetion he promised a high guerdon for
his skill of hand : on this work Eetion has put
forth all his craft.

VIII

Orthon's Grave.

Stranger, the Syracusan Orthon lays this
behest on thee; go never abroad in thy cups
on a night of storm. For thus did I come by
my end, and far from my rich fatherland I lie,
clothed on with alien soil.

IX

The Death of Cleonicus.

Man, husband thy life, nor go voyaging out
of season, for brief are the days of men ! Un-

happy Cleonicus, thou wert eager to win rich
Thasus, from Coelo-Syria sailing with thy merchandise,—with thy merchandise, O Cleonicus,
at the setting of the Pleiades didst thou cross
the sea, — and didst sink with the sinking
Pleiades !

X

A Group of the Muses.

For your delight, all ye Goddesses Nine, did
Xenocles offer this statue of marble, Xenocles
that hath music in his soul, as none will deny.
And inasmuch as for his skill in this art he
wins renown, he forgets not to give their due
to the Muses.

XI

The Grave of Eusthenes.

This is the memorial stone of Eusthenes, the
sage ; a physiognomist was he, and skilled to
read the very spirit in the eyes.   Nobly have
his friends buried him—a stranger in a strange
land—and most dear was he, yea, to the makers
of song.   All his dues in death has the sage,
and, though he was no great one, 'tis plain he
had friends to care for him.

XII

The Offering of Demoteles.

'Twas Demoteles the choregus, O Dionysus,
who dedicated this tripod, and this statue of

thee, the dearest of the blessed gods. No
great fame he won when he gave a chorus of
boys, but with a chorus of men he bore off the
victory, for he knew what was fair and what
was seemly.

XIII

For a statue of Aphrodite.

This is Cypris, — not she of the people;
nay, venerate the goddess by her name—the
Heavenly Aphrodite. The statue is the offer-
ing of chaste Chrysogone, even in the house
of Amphicles, whose children and whose life
were hers! And always year by year went
well with them, who began each year with thy
worship, Lady, for mortals who care for the
Immortals have themselves thereby the better
fortune.

XIV

The Grave of Eurymedon.

An infant son didst thou leave behind, and
in the flower of thine own age didst die, Eurymedon, and win this tomb. For thee a throne
is set among men made perfect, but thy son
the citizens will hold in honour, remembering
the excellence of his father.

XV

The Grave of Eurymedon.

Wayfarer, I shall know whether thou dost
reverence the good, or whether the coward is

held by thee in the same esteem. 'Hail to
this tomb,' thou wilt say, for light it lies above
the holy head of Eurymedon.

XVI

For a statue of Anacreon.

Mark well this statue, stranger, **and** say,
when thou hast returned to thy home, 'In Teos
I beheld the statue of Anacreon, who surely
excelled all the singers of times past.' And if
thou dost add that he delighted in the **young,**
thou wilt truly paint all the man.

XVII

For a statue of Epicharmus.

Dorian is the strain, and Dorian the man
we sing; he that first devised Comedy, even
Epicharmus. O Bacchus, here in bronze (as
the man is now no more) they have erected his
statue, the colonists [1] that dwell in Syracuse, to
the honour of one that was their **fellow-citizen.**
Yea, for a gift he gave, wherefore **we** should be
mindful thereof and pay him what wage we
may, for many maxims he spoke that were
serviceable to the life of all men. Great thanks ,
be his.

[1] Reading πεδοικισται (that is, the Corinthian founders
of Syracuse), and following Wordsworth's other con-
jectures.

XVIII

The Grave of Cleita.

The little Medeus has raised this tomb by the wayside to the memory of his Thracian nurse, and has added the inscription—

HERE LIES CLEITA.

The woman will have this recompense for all her careful nurture of the boy,—and why?—because she was serviceable even to the end.

XIX

The statue of Archilochus.

Stay, and behold Archilochus, him of old time, the maker of the iambics, whose myriad fame has passed westward, alike, and towards the dawning day. Surely the Muses loved him, yea, and the Delian Apollo, so practised and so skilled he grew in forging song, and chanting to the lyre.

XX

The statue of Pisander.

This man, behold, Pisander of Corinth, of all the ancient makers was the first who wrote of the son of Zeus, the lion-slayer, the ready of hand, and spake of all the adventures that with toil he achieved. Know this therefore, that

the people set him here, a statue of bronze,
when many months had gone by and many
years.

XXI

The Grave of Hipponax.

Here lies the poet Hipponax ! If thou art a
sinner **draw not** near this tomb, but if **thou art
a true** man, and the son **of** righteous sires, sit
boldly down here, yea, and sleep if thou wilt.

XXII

For the Bank of Caicus.

To citizens and strangers alike this counter
deals justice. If thou hast deposited aught,
draw out thy money **when the** balance-sheet is
cast up. **Let others make false excuse, but**
Caicus tells back money **lent,** ay, **even if one**
wish it after nightfall.

XXIII

On his own Poems.[1]

The Chian is another **man, but** I, Theocritus,
who wrote these songs, am a Syracusan, **a man**
of the people, being the son of Praxagoras and
renowned Philinna. Never laid I claim to any
Muse but mine **own.**

[1] This epigram **may have** been added by **the first
editor** of Theocritus, **Artemidorus** the Grammarian.

BION

Πίδακος ἐξ ἱερῆς ὀλίγη λιβὰς ἄκρον ἄωτον.—*Callimachus*.

BION was born at Smyrna, one of the towns which claimed the honour of being Homer's birthplace. On the evidence of a detached verse (94) of the dirge by Moschus, some have thought that Theocritus survived Bion. In that **case** Theocritus must **have been a** preternaturally aged man. The same dirge tells us that Bion was poisoned by certain enemies, and that while he left to others his wealth, to Moschus he left his minstrelsy.

BION

THE LAMENT FOR ADONIS

*This poem was probably intended to be sung at one of the
spring celebrations of the festival of Adonis, like that
described by Theocritus in his fifteenth idyl.*

WOE, woe for Adonis, he **hath** perished, **the**
beauteous Adonis, dead is the beauteous Adonis,
the Loves join in the lament.   No more in thy
purple raiment, Cypris, do thou sleep ; arise,
thou wretched one, sable-stoled, and beat thy
breasts, and say to all, 'He hath perished, the
lovely Adonis !'

> *Woe, woe for Adonis, the Loves join in the
> lament !*

Low on the hills is lying the lovely Adonis,
and his thigh with the **boar's** tusk, his white
thigh with the boar's tusk is wounded, and
sorrow on Cypris he brings, as softly he
breathes his life away.

His dark blood drips down his skin of snow,
beneath his brows his eyes wax heavy and dim,
and the rose flees from his lip, and thereon the
very kiss is dying, the kiss that Cypris will
never forego.

To Cypris his kiss is dear, though he lives
no longer, but Adonis knew not that she kissed
him as he died.

> *Woe, woe for Adonis, the Loves join in the lament!*

A cruel, cruel wound on his thigh hath
Adonis, but a deeper wound in her heart
doth Cytherea bear. About him his dear
hounds are loudly baying, and the nymphs of
the wild wood wail him; but Aphrodite with
unbound locks through the glades goes wandering,—wretched, with hair unbraided, with feet
unsandaled, and the thorns as she passes wound
her and pluck the blossom of her sacred blood.
Shrill she wails as down the long woodlands
she is borne, lamenting her Assyrian lord, and
again calling him, and again. But round his
navel the dark blood leapt forth, with blood
from his thighs his chest was scarlet, and
beneath Adonis's breast, the spaces that afore
were snow-white, were purple with blood.

> *Woe, woe for Cytherea, the Loves join in the lament!*

She hath lost her lovely lord, with him she
hath lost her sacred beauty. Fair was the
form of Cypris, while Adonis was living, but

her beauty has died with Adonis! *Woe, woe for Cypris*, the mountains all are saying, and the oak-trees answer, *Woe for Adonis*. And the rivers bewail the sorrows of Aphrodite, and the wells are weeping Adonis on the mountains. The flowers flush red for anguish, and Cytherea through all the mountain-knees, through every dell doth shrill the piteous dirge.

Woe, woe for Cytherea, he hath perished, **the** *lovely Adonis!*

And Echo cried in answer, *He hath perished, the lovely Adonis.* Nay, who but would have lamented the grievous love of Cypris? When she saw, when she marked the unstaunched wound of Adonis, when she saw the bright red blood about his languid thigh, she cast her arms abroad and moaned, 'Abide with me, Adonis, hapless Adonis abide, **that** this last time of all I may possess thee, that I may cast myself about thee, and lips with lips may mingle. Awake Adonis, for a little while, and kiss me yet again, the latest kiss! Nay **kiss me but a moment,** but the lifetime of a kiss, till from thine **inmost soul into my lips,** into my heart, thy life-breath ebb, and till **I** drain thy sweet love-philtre, and drink down all thy love. This kiss will I treasure, even as thyself, Adonis, since, ah ill-fated, thou art fleeing me, thou art fleeing far, Adonis, and art faring to Acheron, to that hateful king and cruel, while wretched I yet live, being a goddess, and may not follow thee! Persephone,

take thou my lover, my lord, for thy self art
stronger than I, and all lovely things drift
down to thee.   But I am all ill-fated, incon-
solable is my anguish, and I lament mine
Adonis, dead to me, and I have no rest for
sorrow.

'Thou diest, O thrice-desired, and my desire
hath flown away as a dream.   Nay, widowed
is Cytherea, and idle are the Loves along the
halls !   With thee has the girdle of my beauty
perished.   For why, ah overbold, didst thou
follow the chase, and being so fair, why wert
thou thus overhardy to fight with beasts ?'

So Cypris bewailed her, the Loves join in
the lament :

> *Woe, woe for Cytherea, he hath perished, the*
> *lovely Adonis !*

A tear the Paphian sheds for each blood-drop
of Adonis, and tears and blood on the earth are
turned to flowers.   The blood brings forth the
rose, the tears, the wind-flower.

> *Woe, woe for Adonis, he hath perished, the*
> *lovely Adonis !*

No more in the oak-woods, Cypris, lament
thy lord.   It is no fair couch for Adonis, the
lonely bed of leaves !   Thine own bed, Cytherea, let him now possess,—the dead Adonis.
Ah, even in death he is beautiful, beautiful in
death, as one that hath fallen on sleep.   Now
lay him down to sleep in his own soft coverlets,
wherein with thee through the night he shared

the holy slumber in a couch all of gold, that
yearns for Adonis, though sad is he to look
upon. Cast on him garlands and blossoms :
all things have perished in his death, yea all
the flowers are faded. Sprinkle him with oint-
ments of Syria, sprinkle him with unguents of
myrrh. Nay, perish all perfumes, for Adonis,
who was thy perfume, hath perished.

He reclines, the delicate Adonis, in his raiment of purple, and around him the Loves are
weeping, and groaning aloud, clipping their
locks for Adonis. And one upon his shafts,
another on his bow is treading, and one hath
loosed the sandal of Adonis, and another hath
broken his own feathered quiver, and one in a
golden vessel bears water, and another laves
the wound, and another from behind him with
his wings is fanning Adonis.

*Woe, woe for Cytherea, the Loves join in the
    lament !*

Every torch on the lintels of the door has
Hymenaeus quenched, and hath torn to shreds
the bridal crown, and *Hymen* no more, *Hymen*
no more is the song, but a new song is sung of
wailing.

' *Woe, woe for Adonis,*' rather than the
nuptial song the Graces are shrilling, lamenting the son of Cinyras, and one to the
other declaring, *He hath perished, the lovely
Adonis.*

And *woe, woe for Adonis*, shrilly cry the
Muses, neglecting Paeon, and they lament

Adonis aloud, and songs they chant to him, but he does not heed them, not that he is loth to hear, but that the Maiden of Hades doth not let him go.

Cease, Cytherea, from thy lamentations, to-day refrain from thy dirges. Thou must again bewail him, again must weep for him another year.

II

THE LOVE OF ACHILLES

Lycidas sings to Myrson a fragment about the loves of Achilles and Deidamia.

Myrson. Wilt thou be pleased now, Lycidas, to sing me sweetly some sweet Sicilian song, some wistful strain delectable, some lay of love, such as the Cyclops Polyphemus sang on the sea-banks to Galatea?

Lycidas. Yes, Myrson, and I too fain would pipe, but what shall I sing?

Myrson. A song of Scyra, Lycidas, is my desire,—a sweet love-story,—the stolen kisses of the son of Peleus, the stolen bed of love; how he, that was a boy, did on the weeds of women, and how he belied his form, and how among the heedless daughters of Lycomedes, Deidamia cherished Achilles in her bower.[1]

---

[1] This conjecture of Meineke's offers, at least, a meaning.

Lycidas. The herdsman bore off Helen, upon a time, and carried her to Ida, sore sorrow to Œnone. And Lacedaemon waxed wroth, and gathered together all the Achaean folk; there was never a Hellene, not one of the Mycenae-ans, nor any man of Elis, nor of the Laconians, that tarried in his house, and shunned the cruel Ares.

But Achilles alone lay hid among the daughters of Lycomedes, and was trained to work in wools, in place of arms, and in his white hand held the bough of maidenhood, in semblance a maiden. For he put on women's ways, like them, and a bloom like theirs blushed on his cheek of snow, and he walked with maiden gait, and covered his locks with the snood. But the heart of a man had he, and the love of a man. From dawn to dark he would sit by Deidamia, and anon would kiss her hand, and oft would lift the beautiful warp of her loom and praise the sweet threads, having no such joy in any other girl of her company. Yea, all things he essayed, and all for one end, that they twain might share an undivided sleep.

Now he once even spake to her, saying—

'With one another other sisters sleep, but I lie alone, and alone, maiden, dost thou lie, both being girls unwedded of like age, both fair, and single both in bed do we sleep. The wicked Nysa, the crafty nurse it is that cruelly severs me from thee. For not of thee have I . . .'

N

III

THE SEASONS

*Cleodamus and Myrson discuss the charms of the seasons,
and give the palm to a southern spring.*

Cleodamus. Which is sweetest, to thee, Myrson, spring, or winter or the late autumn or the summer; of which dost thou most desire the coming? Summer, when all are ended, the toils whereat we labour, or the sweet autumn, when hunger weighs lightest on men, or even idle winter, for even in winter many sit warm by the fire, and are lulled in rest and indolence. Or has beautiful spring more delight for thee? Say, which does thy heart choose? For our leisure lends us time to gossip.

Myrson. It beseems not mortals to judge the works of God; for sacred are all these things, and all are sweet, yet for thy sake I will speak out, Cleodamus, and declare what is sweeter to me than the rest. I would not have summer here, for then the sun doth scorch me, and autumn I would not choose, for the ripe fruits breed disease. The ruinous winter, bearing snow and frost, I dread. But spring, the thrice desirable, be with me the whole year through, when there is neither frost, nor is the sun so heavy upon us. In springtime all is fruitful, all sweet things blossom in spring, and night and dawn are evenly meted to men.

IV

THE BOY AND LOVE

A fowler, while yet a boy, was hunting birds in a woodland glade, and there he saw the winged Love, perched on a box-tree bough. And when he beheld him, he rejoiced, so big the bird seemed to him, and he put together all his rods at once, and lay in wait for Love, that kept hopping, now here, now there. And the boy, being angered that his toil was endless, cast down his fowling gear, and went to the old husbandman, that had taught him his art, and told him all, and showed him Love on his perch. But the old man, smiling, shook his head, and answered the lad, 'Pursue this chase no longer, and go not after this bird. Nay, flee far from him. 'Tis an evil creature. Thou wilt be happy, so long as thou dost not catch him, but if thou comest to the measure of manhood, this bird that flees thee now, and hops away, will come uncalled, and of a sudden, and settle on thy head.'

V

THE TUTOR OF LOVE

Great Cypris stood beside me, while still I slumbered, and with her beautiful hand she led

the child Love, whose head was earthward
bowed. This word she spake to me, 'Dear
herdsman, prithee, take Love, and teach him
to sing.' So said she, and departed, and I—
my store of pastoral song I taught to Love, in
my innocence, as if he had been fain to learn.
I taught him how the cross-flute was invented
by Pan, and the flute by Athene, and by Hermes
the tortoise-shell lyre, and the harp by sweet
Apollo. All these things I taught him as best
I might ; but he, not heeding my words, himself would sing me ditties of love, and taught
me the desires of mortals and immortals, and
all the deeds of his mother. And I clean forgot
the lore I was teaching to Love, but what Love
taught me, and his love ditties, I learned them
all.

VI

LOVE AND THE MUSES

The Muses do not fear the wild Love, but
heartily they cherish, and fleetly follow him.
Yea, and if any man sing that hath a loveless
heart, him do they flee, and do not choose to
teach him. But if the mind of any be swayed
by Love, and sweetly he sings, to him the
Muses all run eagerly. A witness hereto am
I, that this saying is wholly true, for if I sing
of any other, mortal or immortal, then falters
my tongue, and sings no longer as of old, but
if again to Love, and Lycidas I sing, then gladly
from my lips flows forth the voice of song.

FRAGMENTS

VII

I know not the **way,** nor is it fitting to labour at what we have **not** learned.

VIII

If **my** ditties be fair, lo these alone will win me glory, these that the Muse aforetime gave **to me.** And if these be not sweet, what gain **is it** to me to labour longer?

IX

Ah, if a double term of life were given us by Zeus, the son of Cronos, or by changeful Fate, ah, could we spend one life in joy and merriment, and one in labour, then perchance a man might toil, and in some later time might win his reward. But if the gods have willed that man enters into life but once (and that life brief, and too short to hold all we desire), then, wretched **men and weary that we are, how** sorely we toil, **how greatly we** cast **our souls** away on gain, **and laborious** arts, **continually coveting yet more wealth!** Surely we have all forgotten that we are men condemned to die, and how short is the hour, that to us is allotted by Fate.[1]

[1] *Les hommes sont tous condamnés à mort, avec des sursis indéfinis.*—VICTOR HUGO.

X

Happy are they that love, when with equal
love they are rewarded. Happy was Theseus,
when Pirithous was by his side, yea, though
he went down to the house of implacable
Hades. Happy among hard men and inhospitable was Orestes, for that Pylades chose to
share his wanderings. And *he* was happy,
Achilles Æacides, while his darling lived,—
happy was he in his death, because he avenged
the dread fate of Patroclus.

XI

Hesperus, golden lamp of the lovely daughter
of the foam, dear Hesperus, sacred jewel of the
deep blue night, dimmer as much than the
moon, as thou art among the stars pre-eminent,
hail, friend, and as I lead the revel to the
shepherd's hut, in place of the moonlight lend
me thine, for to-day the moon began her
course, and too early she sank. I go not free-
booting, nor to lie in wait for the benighted
traveller, but a lover am I, and 'tis well to
favour lovers.

XII

Mild goddess, in Cyprus born,—thou child,
not of the sea, but of Zeus,—why art thou thus
vexed with mortals and immortals? Nay, my

word is too weak, why wert thou thus bitterly
wroth, yea, even with thyself, as to bring forth
Love, so mighty a bane to all,—cruel and
heartless Love, whose spirit is all unlike his
beauty? And wherefore didst thou furnish him
with wings, and give him skill to shoot so far,
that, child as he is, we never may escape the
bitterness of Love.

XIII

Mute was Phoebus in this grievous anguish.
All herbs he sought, and strove to win some
wise healing art, and he anointed all the
wound with nectar and ambrosia, but remede-
less are all the wounds of Fate.

XIV

But I will go my way to yon sloping hill;
by the sand and the sea-banks murmuring my
song, and praying to the cruel Galatea. But
of my sweet hope never will I leave hold, till I
reach the uttermost limit of old age.

XV

It is not well, my friend, to run to the craftsman, whatever may befall, nor in every matter
to need another's aid, nay, fashion a pipe thyself, and to thee the task is easy.

XVI

May Love call to him the **Muses,** may the
Muses bring with them Love. **Ever** may the
Muses give song to me that yearn **for** it,—
sweet song,—than song there is no **sweeter**
charm.

XVII

The constant dropping of water, **says the**
proverb, it **wears** a hole in a stone.

XVIII

Nay, **leave me not** unrewarded, **for** even
Phoebus sang for **his** reward. **And the** meed
of honour betters everything.

XIX

Beauty is the glory **of** womankind, and
strength **of** men.

XX

All things, god-willing, all things may be
achieved by mortals. From **the** hands **of the**
blessed come **tasks** most **easy, and** that **find**
their accomplishment.

MOSCHUS

OUR only certain information about Moschus is
contained in his own Dirge for Bion. He speaks
of his verse as 'Ausonian song,' and of himself as
Mion's pupil and successor. It is plain that he was
acquainted with the poems of Theocritus.

MOSCHUS

IDYL I

LOVE THE RUNAWAY

CYPRIS was raising the hue and cry for Love,
her child,—'Who, where the three ways meet,
has seen Love wandering? He is my runaway,
whosoever has aught to tell of him shall win
his reward. His prize is the kiss of Cypris,
but if thou bringest him, not the bare kiss, O
stranger, but yet more shalt thou win. The
child is most notable, thou couldst tell him
among twenty together, his skin is not white,
but flame coloured, his eyes are keen and burning, an evil heart and a sweet tongue has he,
for his speech and his mind are at variance.
Like honey is his voice, but his heart of gall,
all tameless is he, and deceitful, the truth is
not in him, a wily brat, and cruel in his pastime.
The locks of his hair are lovely, but his brow is
impudent, and tiny are his little hands, yet far

he shoots his arrows, shoots even to Acheron, and to the King of Hades.

'The body of Love is naked, but well is his spirit hidden, and winged like a bird he flits and descends, now here, now there, upon men and women, and nestles in their inmost hearts. He hath a little bow, and an arrow always on the string, tiny is the shaft, but it carries as high as heaven. A golden quiver on his back he bears, and within it his bitter arrows, wherewith full many a time he wounds even me.

'Cruel are all these instruments of his, but more cruel by far the little torch, his very own, wherewith he lights up the sun himself.

'And if thou catch Love, bind him, and bring him, and have no pity, and if thou see him weeping, take heed lest he give thee the slip ; and if he laugh, hale him along.

'Yea, and if he wish to kiss thee, beware, for evil is his kiss, and his lips enchanted.

'And should he say, "Take these, I give thee in free gift all my armoury," touch not at all his treacherous gifts, for they all are dipped in fire.'

IDYL II

EUROPA AND THE BULL

To Europa, once on a time, a sweet dream
was sent by Cypris, when the third watch of
the night sets in, and near is the dawning;
when sleep more sweet than honey rests on the
eyelids, limb-loosening sleep, that binds the eyes
with his soft bond, when the flock of truthful
dreams fares wandering.

At that hour she was sleeping, beneath the
roof-tree of her home, Europa, the daughter of
Phoenix, being still a maid unwed.   Then she
beheld two Continents at strife for her sake,
Asia, and the farther shore, both in the shape
of women.   Of these one had the guise of a
stranger, the other of a lady of that land, and
closer still she clung about her maiden, and
kept saying how 'she was her mother, and
herself had nursed Europa.'   But that other
with mighty hands, and forcefully, kept haling
the maiden, nothing loth; declaring that, by
the will of Ægis-bearing Zeus, Europa was
destined to be her prize.

But Europa leaped forth from her strown

bed in terror, with beating heart, in such clear
vision had she beheld the dream.   Then she
sat upon her bed, and long was silent, still
beholding the two women, albeit with waking
eyes ; and at last the maiden raised her timor-
ous voice :—

‘ Who of the gods of heaven has sent forth
to me these phantoms ?   What manner of
dreams have scared me when right sweetly
slumbering on my strown bed, within my
bower ? Ah, and who was the alien woman
that I beheld in my sleep ?   How strange a
longing for her seized my heart, yea, and how
graciously she herself did welcome me, and
regard me as it had been her own child.

‘Ye blessed gods, I pray you, prosper the
fulfilment of the dream.’

Therewith she arose, and began to seek the
dear maidens of her company, girls of like age
with herself, born in the same year, beloved of
her heart, the daughters of noble sires, with
whom she was always wont to sport, when she
was arrayed for the dance, or when she would
bathe her bright body at the mouths of the rivers,
or would gather fragrant lilies on the leas.

And soon she found them, each bearing in
her hand a basket to fill with flowers, and to
the meadows near the salt sea they set forth,
where always they were wont to gather in their
company, delighting in the roses, and the
sound of the waves.   But Europa herself bore
a basket of gold, a marvel well worth gazing
on, a choice work of Hephaestus. He gave it

to Libya, for a bridal-gift, when she approached the bed of the Shaker of the Earth, and Libya gave it to beautiful Telephassa, who was of her own blood ; and to Europa, still an unwedded maid, her mother, Telephassa, gave the splendid gift.

Many bright and cunning things were wrought in the basket : therein was Io, daughter of Inachus, fashioned in gold ; still in the shape of a heifer she was, and had not her woman's shape, and wildly wandering she fared upon the salt sea-ways, like one in act to swim ; and the sea was wrought in blue steel. And aloft upon the double brow of the shore, two men were standing together and watching the heifer's sea-faring. There too was Zeus, son of Cronos, lightly touching with his divine hand the cow of the line of Inachus, and her, by Nile of the seven streams, he was changing again, from a horned heifer to a woman. Silver was the stream of Nile, and the heifer of bronze and Zeus himself was fashioned in gold. And all about, beneath the rim of the rounded basket, was the story of Hermes graven, and near him lay stretched out Argus, notable for his sleepless eyes. And from the red blood of Argus was springing a bird that rejoiced in the flower-bright colour of his feathers, and spreading abroad his tail, even as some swift ship on the sea doth spread all canvas, was covering with his plumes the lips of the golden vessel. Even thus was wrought the basket of the lovely Europa.

Now the girls, so soon as they were come to
the flowering meadows, took great delight in
various sorts of flowers, whereof one would
pluck sweet - breathed narcissus, another the
hyacinth, another the violet, a fourth the creeping thyme, and on the ground there fell many
petals of the meadows rich with spring. Others
again were emulously gathering the fragrant
tresses of the yellow crocus ; but in the midst
of them all the princess culled with her hand
the splendour of the crimson rose, and shone
pre-eminent among them all like the foam-born
goddess among the Graces. Verily she was
not for long to set her heart's delight upon the
flowers, nay, nor long to keep untouched her
maiden girdle. For of a truth, the son of
Cronos, so soon as he beheld her, was
troubled, and his heart was subdued by the
sudden shafts of Cypris, who alone can conquer
even Zeus. Therefore, both to avoid the wrath
of jealous Hera, and being eager to beguile the
maiden's tender heart, he concealed his god-
head, and changed his shape, and became a
bull. Not such an one as feeds in the stall
nor such as cleaves the furrow, and drags the
curved plough, nor such as grazes on the grass,
nor such a bull as is subdued beneath the yoke,
and draws the burdened wain. Nay, but while
all the rest of his body was bright chestnut, a
silver circle shone between his brows, and his
eyes gleamed softly, and ever sent forth light-
ning of desire. From his brow branched horns
of even length, like the crescent of the horned

moon, when her disk is cloven in twain. He
came into the meadow, and his coming terrified
not the maidens, nay, within them all wakened
desire to draw nigh the lovely bull, and to
touch him, and his heavenly fragrance was
scattered afar, exceeding even the **sweet per-**
fume of the **meadows.** And he stood before
the feet of fair Europa, and kept licking **her**
neck, and **cast** his spell over the maiden. And
she **still** caressed him, and gently with her
hands she wiped away the deep foam from his
lips, and kissed the bull. Then he lowed so
gently, ye would think ye heard the Mygdonian
flute uttering a dulcet sound.

He bowed himself before her feet, and, bending back his neck, he gazed on Europa, and
showed her his broad back. Then she spake
among her deep-tressed maidens, saying—

'Come, dear playmates, maidens of like **age**
with me, let us mount the bull here and take
our pastime, for truly, he will bear us on his
back, and carry all of us; and how mild he is,
and dear, and gentle to behold, and no whit
like other bulls. A mind as honest as a man's
possesses him, and he lacks nothing but
speech.'

So **she** spake, and smiling, she sat down on
the back of the bull, and the others were about
to follow her. But the bull leaped up immediately, **now** he had gotten her that he
desired, and swiftly he sped to the deep. The
maiden turned, and called again and again to
her dear playmates, stretching out her hands,

O

but they could not reach her. The strand he gained, and forward he sped like a dolphin, faring with unwetted hooves over the wide waves. And the sea, as he came, grew smooth, and the sea-monsters gambolled around, before the feet of Zeus, and the dolphin rejoiced, and rising from the deeps, he tumbled on the swell of the sea. The Nereids arose out of the salt water, and all of them came on in orderly array, riding on the backs of sea-beasts. And himself, the thund'rous Shaker of the World, appeared above the sea, and made smooth the wave, and guided his brother on the salt sea path; and round him were gathered the Tritons, these hoarse trumpeters of the deep, blowing from their long conches a bridal melody.

Meanwhile Europa, riding on the back of the divine bull, with one hand clasped the beast's great horn, and with the other caught up the purple fold of her garment, lest it might trail and be wet in the hoar sea's infinite spray. And her deep robe was swelled out by the winds, like the sail of a ship, and lightly still did waft the maiden onward. But when she was now far off from her own country, and neither sea-beat headland nor steep hill could now be seen, but above, the air, and beneath, the limitless deep, timidly she looked around, and uttered her voice, saying—

'Whither bearest thou me, bull-god? What art thou? how dost thou fare on thy feet through the path of the sea-beasts, nor fearest

the sea ? The sea is a path meet for swift
ships that traverse the brine, but bulls dread
the salt sea-ways. What drink is sweet to
thee, what food shalt thou find from the deep ?
Nay, art thou then some god, for godlike are
these deeds of thine ? Lo, neither do dolphins
of the brine fare on land, nor bulls on the deep,
but dreadless dost thou rush o'er land and sea
alike, thy hooves serving thee for oars.

'Nay, perchance thou wilt rise above the
grey air, and flee on high, like the swift birds.
Alas for me, and alas again, for mine exceeding evil fortune, alas for me that have left my
father's house, and following this bull, on a
strange sea-faring I go, and wander lonely.
But I pray thee that rulest the grey salt sea,
thou Shaker of the Earth, propitious meet me,
and methinks I see thee smoothing this path
of mine before me. For surely it is not without a god to aid, that I pass through these
paths of the waters !'

So spake she, and the horned bull made
answer to her again—

'Take courage, maiden, and dread not the
swell of the deep. Behold I am Zeus, even I,
though, closely beheld, I wear the form of a
bull, for I can put on the semblance of what
thing I will. But 'tis love of thee that has
compelled me to measure out so great a space
of the salt sea, in a bull's shape. Lo, Crete
shall presently receive thee, Crete that was
mine own foster-mother, where thy bridal
chamber shall be. Yea, and from me shalt

thou bear glorious sons, to be sceptre-swaying kings over earthly men.'

So spake he, and all he spake was fulfilled. And verily Crete appeared, and Zeus took his own shape again, and he loosed her girdle, and the Hours arrayed their bridal bed. She that before was a maiden straightway became the bride of Zeus, and she bare children to Zeus, yea, anon she was a mother.

THE LAMENT FOR BION

WAIL, let me hear you wail, ye woodland glades, and thou Dorian water; and weep ye rivers, for Bion, the well beloved! Now all ye green things mourn, and now ye groves lament him, ye flowers now in sad clusters breathe yourselves away. Now redden ye roses in your sorrow, and now wax red ye wind-flowers, now thou hyacinth, whisper the letters on thee graven, and add a deeper *ai ai* to thy petals; he is dead, the beautiful singer.

Begin, ye Sicilian Muses, begin the dirge.

Ye nightingales that lament among the thick leaves of the trees, tell ye to the Sicilian waters of Arethusa the tidings that Bion the herdsman is dead, and that with Bion song too has died, and perished hath the Dorian minstrelsy.

Begin, ye Sicilian Muses, begin the dirge.

Ye Strymonian swans, sadly wail ye by the waters, and chant with melancholy notes the dolorous song, even such a song as in his time

with voice like yours he was wont to sing.
And tell again to the Œagrian maidens, tell to
all the Nymphs Bistonian, how that he hath
perished, the Dorian Orpheus.

Begin, ye Sicilian Muses, begin the dirge.

No more to his herds he sings, that beloved
herdsman, no more 'neath the lonely oaks he
sits and sings, nay, but by Pluteus's side he
chants a refrain of oblivion. The mountains
too are voiceless: and the heifers that wander
by the bulls lament and refuse their pasture.

Begin, ye Sicilian Muses, begin the dirge.

Thy sudden doom, O Bion, Apollo himself
lamented, and the Satyrs mourned thee, and
the Priapi in sable raiment, and the Panes
sorrow for thy song, and the fountain fairies in
the wood made moan, and their tears turned
to rivers of waters. And Echo in the rocks
laments that thou art silent, and no more she
mimics thy voice. And in sorrow for thy fall
the trees cast down their fruit, and all the
flowers have faded. From the ewes hath
flowed no fair milk, nor honey from the hives,
nay, it hath perished for mere sorrow in the
wax, for now hath thy honey perished, and no
more it behoves men to gather the honey of
the bees.

Begin, ye Sicilian Muses, begin the dirge.

Not so much did the dolphin mourn beside
the sea-banks, nor ever sang so sweet the
nightingale on the cliffs, nor so much lamented

the swallow on the long ranges of the hills, nor
shrilled so loud the halcyon o'er his sorrows ;
   (*Begin, ye Sicilian Muses, begin the dirge.*)

Nor so much, by the grey sea-waves, did
ever the sea-bird sing, nor so much in the dells
of dawn did the bird of Memnon bewail the
son of the Morning, fluttering around his tomb,
as they lamented for Bion dead.

   Nightingales, and all the swallows that once
he was wont to delight, that he would teach
to speak, they sat over against each other on
the boughs and kept moaning, and the birds
sang in answer, ' Wail, ye wretched ones, even
ye ! '
   Begin, ye Sicilian Muses, begin the dirge.

Who, ah who will ever make music on thy
pipe, O thrice desired Bion, and who will put
his mouth to the reeds of thine instrument ?
who is so bold ?

   For still thy lips and still thy breath survive,
and Echo, among the reeds, doth still feed
upon thy songs.   To Pan shall I bear the
pipe ?   Nay, perchance even he would fear to
set his mouth to it, lest, after thee, he should
win but the second prize.
   Begin, ye Sicilian Muses, begin the dirge.

Yea, and Galatea laments thy song, she
whom once thou wouldst delight, as with thee
she sat by the sea-banks.   For not like the
Cyclops didst thou sing—him fair Galatea ever
fled, but on thee she still looked more kindly

than on the salt water. And now hath she forgotten the wave, and sits on the lonely sands,
but still she keeps thy kine.

Begin, ye Sicilian Muses, begin the dirge.

All the gifts of the Muses, herdsman, have
died with thee, the delightful kisses of maidens,
the lips of boys ; and woful round thy tomb the
loves are weeping. But Cypris loves thee far
more than the kiss wherewith she kissed the
dying Adonis.

Begin, ye Sicilian Muses, begin the dirge.

This, O most musical of rivers, is thy second
sorrow, this, Meles, thy new woe. Of old
didst thou lose Homer, that sweet mouth of
Calliope, and men say thou didst bewail thy
goodly son with streams of many tears, and
didst fill all the salt sea with the voice of thy
lamentation — now again another son thou
weepest, and in a new sorrow art thou wasting
away.

Begin, ye Sicilian Muses, begin the dirge.

Both were beloved of the fountains, and one
ever drank of the Pegasean fount, but the other
would drain a draught of Arethusa. And the
one sang the fair daughter of Tyndarus, and
the mighty son of Thetis, and Menelaus
Atreus's son, but that other,—not of wars, not of
tears, but of Pan, would he sing, and of herdsmen would he chant, and so singing, he tended
the herds. And pipes he would fashion, and
would milk the sweet heifer, and taught lads

how to kiss, and Love he cherished in his
bosom and woke the passion of Aphrodite.
Begin, ye Sicilian Muses, begin the dirge.

Every famous city laments thee, Bion, and
all the towns. Ascra laments thee far more
than her Hesiod, and Pindar is less regretted
by the forests of Boeotia. Nor so much did
pleasant Lesbos mourn for Alcaeus, nor did the
Teian town so greatly bewail her poet, while for
thee more than for Archilochus doth Paros
yearn, and not for Sappho, but still for thee
doth Mytilene wail her musical lament ;

[*Here seven verses are lost.*]

And in Syracuse Theocritus ; but I sing thee
the dirge of an Ausonian sorrow, I that am no
stranger to the pastoral song, but heir of the
Doric Muse which thou didst teach thy pupils.
This was thy gift to me ; to others didst thou
leave thy wealth, to me thy minstrelsy.
Begin, ye Sicilian Muses, begin the dirge.

Ah me, when the mallows wither in the
garden, and the green parsley, and the curled
tendrils of the anise, on a later day they live
again, and spring in another year ; but we men,
we, the great and mighty, or wise, when once
we have died, in hollow earth we sleep, gone
down into silence ; a right long, and endless,
and unawakening sleep. And thou too, in the
earth wilt be lapped in silence, but the nymphs
have thought good that the frog should eter-

nally sing. Nay, him I would not envy, for 'tis
no sweet song he singeth.

Begin, ye Sicilian Muses, begin the dirge.

Poison came, Bion, to thy mouth, thou didst
know poison. To such lips as thine did it
come, and was not sweetened? What mortal
was so cruel that could mix poison for thee, or
who could give thee the venom that heard thy
voice? surely he had no music in his soul.

Begin, ye Sicilian Muses, begin the dirge.

But justice hath overtaken them all. Still
for this sorrow I weep, and bewail thy ruin.
But ah, if I might have gone down like Orpheus
to Tartarus, or as once Odysseus, or Alcides of
yore, I too would speedily have come to the
house of Pluteus, that thee perchance I might
behold, and if thou singest to Pluteus, that I
might hear what is thy song. Nay, sing to the
Maiden some strain of Sicily, sing some sweet
pastoral lay.

And she too is Sicilian, and on the shores by
Aetna she was wont to play, and she knew the
Dorian strain. Not unrewarded will the singing be ; and as once to Orpheus's sweet minstrelsy she gave Eurydice to return with him,
even so will she send thee too, Bion, to the
hills. But if I, even I, and my piping had
aught availed, before Pluteus I too would have
sung.

IDYL IV

A sad dialogue between Megara the wife and Alcmena the mother of the wandering Heracles. Megara had seen her own children slain by her lord, in his frenzy, while Alcmena was constantly disquieted by ominous dreams.

MY mother, wherefore art thou thus smitten in thy soul with exceeding sorrow, and the rose is no longer firm in thy cheeks as of yore? why, tell me, art thou thus disquieted? Is it because thy glorious **son** is suffering pains unnumbered in bondage to a man of naught, as it were a lion in bondage to a fawn? Woe is me, why, ah why have the immortal gods thus brought on me so great dishonour, and wherefore did my parents get me for so ill a doom? Wretched woman that I am, who came to the bed of a man without reproach and ever held him honourable and dear as mine own eyes,—ay and still worship and hold him sacred in my heart—yet none other of men living hath had more evil hap or tasted in his soul so many griefs. In madness once, with the bow Apollo's self had given him—dread weapon of some Fury or spirit of Death—he struck down

his own children, and took their dear life away,
as his frenzy raged through the house till it
swam in blood. With mine own eyes, I saw
them smitten, woe is me, by their father's
arrows—a thing none else hath suffered even in
dreams. Nor could I aid them as they cried
ever on their mother ; the evil that was upon
them was past help. As a bird mourneth for
her perishing little ones, devoured in the thicket
by some terrible serpent while as yet they are
fledglings, and the kind mother flutters round
them making most shrill lament, but cannot
help her nestlings, yea, and herself hath great
fear to approach the cruel monster ; so I unhappy mother, wailing for my brood, with
frenzied feet went wandering through the house.
Would that by my children's side I had died
myself, and were lying with the envenomed
arrow through my heart. Would that this had
been, O Artemis, thou that art queen chief of
power to womankind. Then would our parents
have embraced and wept for us and with ample
obsequies have laid us on one common pyre,
and have gathered the bones of all of us into
one golden urn, and buried them in the place
where first we came to be. But now they
dwell in Thebes, fair nurse of youth, ploughing
the deep soil of the Aonian plain, while I in
Tiryns, rocky city of Hera, am ever thus
wounded at heart with many sorrows, nor is any
respite to me from tears. My husband I behold but a little time in our house, for he hath
many labours at his hand, whereat he laboureth

in wanderings by land and sea, with his soul
strong as rock or steel within his breast. But
thy grief is as the running waters, as thou
lamentest through the nights and all the days
of Zeus.

Nor is there **any one of** my kinsfolk nigh at
hand to **cheer me : for it is not the house wall**
that severs **them,** but they all dwell **far** beyond
the pine-clad Isthmus, nor is there any to whom,
as a woman all hapless, I may look up and
refresh my heart, save only my sister Pyrrha ;
nay, but she herself grieves yet more for her
husband Iphicles thy son : for methinks 'tis
thou that hast borne the most luckless children
of all, to a God, and a mortal man.[1]

Thus spake she, and ever warmer the tears
were pouring from her eyes into her sweet
bosom, as she bethought her of her children
and next of her own parents. And in like
manner Alcmena bedewed her pale cheeks with
tears, and deeply sighing from her very heart
she thus bespoke her dear daughter with thick-
coming words :

'Dear child, what is this that hath come
into the thoughts of thy heart ? How art thou
fain to disquiet us both with the tale of griefs
that cannot be forgotten ? Not for the first
time are these woes wept for now. Are they
not enough, the woes that possess us from our
birth continually to our day of death ? In love
with sorrow surely would he be that should

---

[1] Alcmena bore Iphicles to Amphictyon, Hercules to
Zeus.

have the heart to count up our woes; such
destiny have we received from God. Thyself,
dear child, I behold vext by endless pains, and
thy grief I can pardon, yea, for even of joy
there is satiety. And exceedingly do I mourn
over and pity thee, for that thou hast partaken
of our cruel lot, the burden whereof is hung
above our heads. For so witness Persephone
and fair-robed Demeter (by whom the enemy
that wilfully forswears himself, lies to his own
hurt), that I love thee no less in my heart than
if thou hadst been born of my womb, and
wert the maiden darling of my house : nay, and
methinks that thou knowest this well. Therefore say never, my flower, that I heed thee not,
not even though I wail more ceaselessly than
Niobe of the lovely locks. No shame it is for
a mother to make moan for the affliction of her
son : for ten months I went heavily, even
before I saw him, while I bare him under my
girdle, and he brought me near the gates of the
warden of Hell ; so fierce the pangs I endured
in my sore travail of him. And now my son is
gone from me in a strange land to accomplish
some new labour; nor know I in my sorrow
whether I shall again receive him returning
here or no. Moreover in sweet sleep a dreadful dream hath fluttered me ; and I exceedingly
fear for the ill-omened vision that I have seen,
lest something that I would not be coming on
my children.

It seemed to me that my son, the might of
Heracles, held in both hands a well-wrought

spade, wherewith, as one labouring for hire, he
was digging a ditch at the edge of a fruitful
field, stripped of his cloak and belted tunic.
And when he had come to the end of all his
work and his labours at the stout defence of the
vine-filled close, he was about to lean his shovel
against the upstanding mound and don the
clothes he had worn.   But suddenly blazed up
above the deep trench a quenchless fire, and a
marvellous great flame encompassed him.   But
he kept ever giving back with hurried feet,
striving to flee the deadly bolt of Hephaestus;
and ever before his body he kept his spade as
it were a shield; and this way and that he
glared around him with his eyes, lest the angry
fire should consume him.   Then brave Iphicles,
eager, methought, to help him, stumbled and
fell to earth ere he might reach him, nor could
he stand upright again, but lay helpless, like a
weak old man, whom joyless age constrains to
fall when he would not; so he lieth on the
ground as he fell, till one passing by lift him
up by the hand, regarding the ancient reverence
for his hoary beard.   Thus lay on the earth
Iphicles, wielder of the shield.   But I kept
wailing as I beheld my sons in their sore
plight, until deep sleep quite fled from my
eyes, and straightway came bright morn.   Such
dreams, beloved, flitted through my mind all
night; may they all turn against Eurystheus
nor come nigh our dwelling, and to his hurt be
my soul prophetic, nor may fate bring aught
otherwise to pass.

IDYL V

WHEN the wind on the grey salt sea blows
softly, then my weary spirits rise, and the land
no longer pleases me, and far more doth the
calm allure me.[1] But when the hoary deep is
roaring, and the **sea** is broken up in foam, and
the waves rage high, then lift I mine eyes unto
the earth and trees, and fly **the sea**, and the
land **is** welcome, **and the shady** wood well
pleasing in my sight, where even if the wind
blow high the pine-tree sings her song.   Surely
an evil life lives the fisherman, whose home is
his ship, and his labours are in the sea, and
fishes thereof are his wandering spoil.   Nay,
sweet to me is sleep beneath the broad-leaved
plane-tree ; let me love to listen to the murmur
of the brook hard by, soothing, not troubling
the husbandman with its sound.

IDYL VI

PAN loved his neighbour Echo ; Echo loved
A gamesome Satyr ; he, by her unmoved,

[1] Reading, with Weise, ποτάγει δὲ πολὺ πλεόν ἄμμε
γαλάνα.

Loved only Lyde ; thus through Echo, Pan,
Lyde, and Satyr, Love his circle ran.
Thus all, while their **true** lovers' hearts they
    grieved,
Were scorned **in turn, and** what they **gave**
    received.
O all Love's scorners, learn this lesson true ;
Be kind **to Love, that he be kind to you.**

IDYL VII

ALPHEUS, when he leaves Pisa and makes his
way through beneath the deep, travels on to
Arethusa with his waters that the wild olives
drank, bearing her bridal gifts, fair leaves **and**
flowers and sacred soil.   Deep in the waves he
plunges, and runs beneath the sea, and the salt
water mingles **not with the sweet.**   Nought
knows the **sea as the river** journeys **through.**
Thus hath **the knavish boy, the** maker **of mis-**
chief, the **teacher** of strange ways—thus hath
Love by his spell **taught even a** river to dive.

IDYL VIII

LEAVING his torch and his arrows, **a** wallet
    strung on his back,
One day came the mischievous Love-god **to**
    follow the plough-share's track :
And he **chose** him **a staff** for his driving, and
    yoked him a sturdy steer,

P

And sowed in the furrows the grain to the
    Mother of Earth most dear.
Then he said, looking up to the **sky** : ' Father
    Zeus, to my harvest be good,
Lest I yoke that bull to my plough that Europa
    once **rode** through the flood ! '

IDYL IX

WOULD that **my** father **had taught** me the craft
    of a keeper of sheep,
For so in the shade of the elm-tree, or under
    the rocks on the steep,
Piping **on reeds I had sat, and** had lulled my
    sorrow to sleep.[1]

---

[1] For the translations into verse I have to thank Mr.
Ernest Myers.

THE END

Printed by R. & R. CLARK, *Edinburgh.*